SpringerBriefs in Statistics

JSS Research Series in Statistics

The current research of statistics in Japan has expanded in several directions in line with recent trends in academic activities in the area of statistics and statistical sciences over the globe. The core of these research activities in statistics in Japan has been the Japan Statistical Society (JSS). This society, the oldest and largest academic organization for statistics in Japan, was founded in 1931 by a handful of pioneer statisticians and economists and now has a history of about 80 years. Many distinguished scholars have been members, including the influential statistician Hirotugu Akaike, who was a past president of JSS, and the notable mathematician Kiyosi Itô, who was an earlier member of the Institute of Statistical Mathematics (ISM), which has been a closely related organization since the establishment of ISM. The society has two academic journals: the Journal of the Japan Statistical Society (English Series) and the Journal of the Japan Statistical Society (Japanese Series). The membership of JSS consists of researchers, teachers, and professional statisticians in many different fields including mathematics, statistics, engineering, medical sciences, government statistics, economics, business, psychology, education, and many other natural, biological, and social sciences. The JSS Series of Statistics aims to publish recent results of current research activities in the areas of statistics and statistical sciences in Japan that otherwise would not be available in English; they are complementary to the two JSS academic journals, both English and Japanese. Because the scope of a research paper in academic journals inevitably has become narrowly focused and condensed in recent years, this series is intended to fill the gap between academic research activities and the form of a single academic paper. The series will be of great interest to a wide audience of researchers, teachers, professional statisticians, and graduate students in many countries who are interested in statistics and statistical sciences, in statistical theory, and in various areas of statistical applications.

More information about this subseries at http://www.springer.com/series/13497

Yasunori Fujikoshi · Vladimir V. Ulyanov

Non-Asymptotic Analysis of Approximations for Multivariate Statistics

Springer

Yasunori Fujikoshi
Hiroshima University
Higashi-Hiroshima, Japan

Vladimir V. Ulyanov
National Research University
Higher School of Economics
Moscow State University
Moscow, Russia

ISSN 2191-544X ISSN 2191-5458 (electronic)
SpringerBriefs in Statistics
ISSN 2364-0057 ISSN 2364-0065 (electronic)
JSS Research Series in Statistics
ISBN 978-981-13-2615-8 ISBN 978-981-13-2616-5 (eBook)
https://doi.org/10.1007/978-981-13-2616-5

This Springer imprint is published by the registered company Springer Nature Singapore Pte Ltd.
The registered company address is: 152 Beach Road, #21-01/04 Gateway East, Singapore 189721, Singapore

Preface

This book provides readers with recent non-asymptotic results for approximations in multivariate statistical analysis. There are many traditional multivariate methods based on large-sample approximations. Furthermore, in recent years more high-dimensional multivariate methods have been proposed and utilized for cases where the dimension p of observations is comparable with the sample size n or even exceeds it. Related to this, there are also many approximations under high-dimensional frameworks when $p/n \to c \in (0, 1)$ or $(0, \infty)$.

An important problem related to multivariate approximations concerns their errors. Most results contain only so-called order estimates. However, such error estimates do not provide information on actual errors for given values of n, p, and other parameters. Ideally, we need non-asymptotic or computable error bounds that relate to these actual errors, in addition to order estimates. In non-asymptotic bounds, the pair (n, p), as well as other problem parameters, are viewed as fixed, and statistical statements such as tail or concentration probabilities of test statistics and estimators are constructed as a function of them. In other words, these results are applied for actual values of (n, p). In general, non-asymptotic error bounds involve an absolute constant. If the absolute constant is known, then such an error bound is called the computable error bound.

Our book focuses on non-asymptotic bounds for high-dimensional and large-sample approximations. A brief explanation of non-asymptotic bounds is given in Chap. 1. Some commonly used notations are also explained in Chap. 1. Chapters 2–6 deal with computable error bounds. In Chap. 2, the authors consider computable error bounds on scale-mixed variables. The results can be applied to asymptotic approximations of t- and F-distributions, and to various estimators. In Chap. 3, error bounds for MANOVA tests are given based on large-sample results for multivariate scale mixtures. High-dimensional results are also given. In Chap. 4, the focus is on linear and quadratic discriminant contexts, with error bounds for location and scale mixture variables. In Chaps. 5 and 6, computable error bounds for Cornish–Fisher expansions and Λ-statistics are considered, respectively.

Next, in Chaps. 7–11, new directions of research on non-asymptotic bounds are discussed. In Chap. 7, the focus is on high-dimensional approximations for bootstrap procedures in principal component analysis. Then, in Chap. 8 we consider the Kolmogorov distance between the probabilities of two Gaussian elements to hit a ball in Hilbert space. In Chap. 9, the focus is on approximations of statistics based on observations with random sample sizes. In Chap. 10, the topic is large-sample approximations of power-divergence statistics including the Pearson chi-squared statistic, the Freeman–Tukey statistics, and the log-likelihood ratio statistic. Finally, Chap. 11 proposes a general approach for constructing non-asymptotic estimates and provides relevant examples for several complex statistics.

This book is intended to be used as a reference for researchers interested in asymptotic approximations in multivariate statistical analysis contexts. It will also be useful for instructors and students of graduate-level courses as it covers important foundations and methods of multivariate analysis.

For many approximations, detailed derivations would require a lot of space. For the sake of brevity and presentation, we therefore mainly give their outline. We believe and hope that the book will be useful for stimulating future developments in non-asymptotic analysis of multivariate approximations.

We are very grateful to our colleagues R. Shimizu, F. Götze, G. Christoph, V. Spokoiny, H. Wakaki and A. Naumov to be our co-authors for years. Our joint works are widely used in the book. We express our sincere gratitude to Prof. Naoto Kunitomo, Meiji University, Tokyo, and Dr. Tetsuro Sakurai, Suwa University of Science, Nagano, for their valuable comments on various aspects of the content of this book. We are also grateful to Mr. Y. Hirachi for his assistance in the preparation of this book.

The research was partially supported by the Ministry of Education, Science, Sports, and Culture through a Grant-in-Aid for Scientific Research (C), 16K00047, 2016–2018, and results of Chapters 8 and 9 have been obtained under support of the RSF Grant No. 18-11-00132. The study was done within the framework of the Moscow Center for Fundamental and Applied Mathematics, Moscow State University and HSE University Basic Research Programs.

Hiroshima, Japan
Moscow, Russia
April 2020

Yasunori Fujikoshi
Vladimir V. Ulyanov

Contents

Chapter 1
Non-Asymptotic Bounds

Abstract Most asymptotic errors in statistical inference are based on error estimates when the sample size n and the dimension p of observations are large. More precisely, such statistical statements are evaluated when n and/or p tend to infinity. On the other hand, "non-asymptotic" results are derived under the condition that n, p, and the parameters involved are fixed. In this chapter, we explain non-asymptotic error bounds, while giving the Edgeworth expansion, Berry–Essen bounds, and high-dimensional approximations for the linear discriminant function.

1.1 Errors and Non-Asymptotic Bounds

In mathematical statistics and probability theory, asymptotics are used in analysis of long-run or large-sample and high-dimensional behavior of random variables related to various statistical inference contexts. Most asymptotic theory is based on results when the sample size n and the dimension p of observations tend to infinity. Often, such results do not provide information that is relevant in situations where the sample size and dimension are given finite values. To appreciate this more precisely, consider Edgeworth-type expansion. Let $\{F_n(x)\}$ be a set of distribution functions indexed by a parameter n, typically, sample size. For example, suppose $F_n(x)$ is approximated by the first k terms of asymptotic expansion of the Edgeworth type:

$$G_{k,n}(x) = G(x) + \sum_{j=1}^{k-1} n^{-j/2} p_j(x) g(x), \tag{1.1}$$

where $G(x)$ is the limiting distribution function of $F_n(x)$, $g(x)$ is the pdf of $G(x)$, and the $p_j(x)$'s are suitable polynomials. For a wide class of statistics, it is known (see, e.g., [1]) that the error $R_{k,n}(x) = F_n(x) - G_{k,n}(x)$ satisfies the order condition

$$R_{k,n}(x) = \mathrm{O}(n^{-k/2}), \quad \text{uniformly in } x. \tag{1.2}$$

Y. Fujikoshi and V. V. Ulyanov, *Non-Asymptotic Analysis of Approximations for Multivariate Statistics*, JSS Research Series in Statistics,
https://doi.org/10.1007/978-981-13-2616-5_1

Note that this statement is rather implicit because it does not give any information about the largest possible value of $R_{k,n}(x)$ for a given n. More precisely, condition (1.2) means that there exists a positive constant C_k and a positive number N_k such that the inequality

$$\sup_x |R_{k,n}(x)| \le C_k / n^{k/2} \quad \text{for all } n \ge N_k \tag{1.3}$$

holds. However, the values of C_k and N_k are unknown. In some cases, the situation becomes even worse: F_n may depend on a "nuisance" parameter, say, $\theta \in \Theta$. Then C_k and N_k depend on θ as well, i.e., $C_k = C_k(\theta)$ and $N_k = N_k(\theta)$, which are, again, totally unknown except that they are finite, or at best, that they are monotone in θ.

Our knowledge about the error bound will improve if we can fix a number C_k for a given N_k, possibly depending on θ, in such a way that inequality (1.3) is true. We call this kind of error bound a computable error bound or a non-asymptotic error bound. In general, the term "computable error bound" means that the error can be numerically computed when (n, p) and the values of parameters are given.

One of the main problems associated with approximations in multivariate statistical analysis concerns their errors. Most results only supply so-called order estimates. Such estimates do not provide information on actual errors for given values of sample size n, dimension p, and other parameters. It is desirable to develop non-asymptotic results whose statements also hold for fixed values of (n, p) and other problem parameters. Therefore, ideally, we want to have non-asymptotic bounds or computable error bounds, in addition to order estimates.

The most well-known computable error bound is the Berry–Esseen theorem, or the bound in the central limit theorem. For expository purposes we consider the case when $X_1, X_2, \ldots$ are i.i.d. random variables. Without loss of generality, we assume that $\mathrm{E}(X_1) = 0$ and $\mathrm{Var}(X_1) = \mathrm{E}(X_1^2) = 1$. Let $\beta_3 = \mathrm{E}(|X_1|^3)$ be finite and let $F_n(x)$ and $\Phi(x)$ be the distribution functions of the normalized sum $(X_1 + \cdots + X_n)/\sqrt{n}$ and the standard normal distribution, respectively. Then, it is known that there exists a positive constant C such that for all n

$$\sup_x \left| F_n(x) - \Phi(x) \right| \le C\beta_3/\sqrt{n}. \tag{1.4}$$

There are many works on seeking a value of C. It is known [5] that $C \le 0.4748$ for all $n \ge 1$. In terms of error bounds for its first-order expansion, see [2], for example. However, these topics are beyond the scope of this book.

As another non-asymptotic result, consider misclassification errors in two-group discriminant analysis. Suppose we are interested in classifying a $p \times 1$ observation vector $\boldsymbol{X}$ as coming from one of two populations Π_1 or Π_2. Let Π_i be a two p-variate normal population with $\mathrm{N}_p(\boldsymbol{\mu}_i, \boldsymbol{\Sigma})$, where $\boldsymbol{\mu}_1 \ne \boldsymbol{\mu}_2$ and $\boldsymbol{\Sigma}$ is positive definite. When the values of the parameters are unknown, assume that random samples of sizes n_1 and n_2 are available from Π_1 and Π_2, respectively. Let $\bar{\boldsymbol{X}}_1$, $\bar{\boldsymbol{X}}_2$, and $\mathbf{S}$ be the sample mean vectors and the sample covariance matrix. Then, a well-known linear discriminant function is defined by $W = (\bar{\boldsymbol{X}}_1 - \bar{\boldsymbol{X}}_2)'\mathbf{S}^{-1}\left\{\boldsymbol{X} - \frac{1}{2}(\bar{\boldsymbol{X}}_1 + \bar{\boldsymbol{X}}_2)\right\}$. The observation

X may be classified to Π_1 or Π_2 according to $W \geq 0$ or $W < 0$. We are interested in the misclassification probabilities given by $e_W(2|1) = \mathrm{P}(W \leq 0 \mid X \in \Pi_1)$ and $e_W(1|2) = \mathrm{P}(W \leq 0 \mid X \in \Pi_2)$. Under a high-dimensional asymptotic framework $n_1 \to \infty, n_2 \to \infty, m = n_1 + n_2 - p \to \infty$, it is known (see, e.g., [3, 4]) that $e_W(2|1)$ converges to $\Phi(\gamma)$, and

$$|e_W(2|1) - \Phi(\gamma)| \leq B(p, n_1, n_2, \Delta^2), \tag{1.5}$$

where $\Delta^2 = (\boldsymbol{\mu}_1 - \boldsymbol{\mu}_2)'\boldsymbol{\Sigma}^{-1}(\boldsymbol{\mu}_1 - \boldsymbol{\mu}_2)$ is the square of the Mahalanobis distance. For closed forms of γ and $B(p, n_1, n_2, \Delta^2)$, see [4]. Similar results have been obtained for $e_W(1|2)$. Note that result (1.5) holds for any given (p, n_1, n_2, Δ^2) such that $m > 7$; thus it is a non-asymptotic result. Further, $B(p, n_1, n_2, \Delta^2)$ is a computable error bound of order $\mathrm{O}(m^{-1})$. Similar results are given for the quadratic discriminant function. In Chap. 4, we show such results by extending non-asymptotic theory for a location and scale mixture.

In general, approximation errors can be described either asymptotically as an order of a remainder term with respect to a sample size n and/or a dimension p, or non-asymptotically as a bound for a remainder term with clearly expressed dependence on n, moment characteristics, and p. In this book, we deal with the latter non-asymptotic results and try to determine the correct structure of the bound for a remainder term in relation to all parameters involved, with the exception of the absolute constants. If we can give the values of absolute constants as well, then the corresponding bounds are called computable. Such computable bounds were considered for some multivariate statistics in the earlier book by [4]. In this new book, first we consider computable error bounds on scale-mixed variables, MANOVA tests, discriminant functions, Cornish–Fisher expansions, and some other statistics. These considerations are based on the recent findings, where applicable.

After the book by Fujikoshi et al. [4] was published by Wiley in 2010, a series of new non-asymptotic results appeared. Research continued in new directions, in particular, for bootstrap procedures, for approximations of statistics based on observations with random sample sizes, and for power-divergence statistics. These results are presented in this book. Further, in Chap. 11, we also suggest a general approach for constructing non-asymptotic bounds and provide corresponding examples for relatively complex statistics.

In statistics, there are many asymptotic results including asymptotic expansions as n and/or p tend to infinity. Some of them are given without rigorous proofs for their error terms. Non-asymptotic results will also be useful as rigorous proofs for the order estimates of such approximation errors.

Here, we note that some commonly used notations are used in this book without defining them in detail. For example, the transpose of a matrix $\mathbf{A}$ is denoted by $\mathbf{A}'$ or $\mathbf{A}^{\mathsf{T}}$. For a squared matrix $\mathbf{A}$, its determinant and trace are denoted by $|\mathbf{A}|$ and $\mathrm{tr}\mathbf{A}$, respectively. The usual norm $(\boldsymbol{a}'\boldsymbol{a})^{1/2}$ of a vector $\boldsymbol{a}$ is denoted by $\|\boldsymbol{a}\|$ or $\|\boldsymbol{a}\|_2$. For a random vector $\boldsymbol{X}$, its mean is denoted by $\mathrm{E}(\boldsymbol{X})$. The covariance matrix is denoted by $\mathrm{Var}(\boldsymbol{X})$ or $\mathrm{Cov}(\boldsymbol{X})$.

References

1. Bhattacharya, R. N., & Ghosh, J. K. (1978). On the validity of the formal Edgeworth expansion. *The Annals of Statistics*, *6*, 434–451. (corrigendum ibid., **8**, 1980).
2. Dobrić, V., & Ghosh, B. K. (1996). Some analogs of the Berry-Esseen bound for first-order Chebyshev–Edgeworth expansions. *Statistics & Decision*, *14*, 383–404.
3. Fujikoshi, Y. (2000). Error bounds for asymptotic approximations of the linear discriminant function when the sample size and dimensionality are large. *Journal of Multivariate Analysis*, *73*, 1–17.
4. Fujikoshi, Y., Ulyanov, V. V., & Shimizu, R. (2010). *Multivariate Analysis: High-Dimensional and Large-Sample Approximations*. Hoboken: Wiley.
5. Shevtsova, I. G. (2011). On the absolute constants in the Berry–Esseen type inequalities for identically distributed summands. arXiv: 1111.6554.

Chapter 2
Scale-Mixed Distributions

Abstract In this chapter we present a general theory of approximation of scale-mixed distributions or distributions of scale mixtures, including simple examples of Student's t-distribution and F-distribution as a scale mixtures of the normal and chi-square distribution, respectively. Such scale mixtures appear as sampling distributions of various statistics such as the studentized version of some estimators. Errors of the approximation are evaluated in sup and L_1-norms. Extension to multivariate scale mixtures with error bounds evaluated in L_1-norm shall be discussed in Chap. 3.

2.1 Introduction

The purpose of the present chapter is to develop a theory of approximating the distribution of the scale mixture of some underlying distribution G, i.e., the distribution of

$$X = S^{\delta} Z, \tag{2.1}$$

where Z has a distribution function G, while S is a positive random variable independent of Z and $\delta = \pm 1$ is a parameter used to distinguish two types of mixtures: $X = SZ$ and $X = S^{-1}Z$.

A relatively wide class of sampling distributions can be expressed as a mixture of the standard normal or chi-square distribution. For example, when Z is distributed to $\mathrm{N}(0, 1)$, $S = \sqrt{\chi_n^2/n}$ and $\delta = -1$, then, X is distributed to t-distribution with n degrees of freedom. When Z is distributed to a χ_q^2-distribution with q degrees of freedom, $S = \chi_n^2/n$ and $\delta = -1$, then, X/q is distributed to F-distribution with (q, n) degrees of freedom. We show that the scale mixtures with $\delta = 1$ have important applications.

For another example, we have a studentized statistic $V^{-1}Z$ of the form (2.1) with $\delta = -1$, $Z = U/\sigma$, and $S = V/\sigma$ (where σ is an unknown population parameter). Here, Z has a known distribution G and V^2 is an estimator of σ^2 which is independent of Z. Naturally, the distribution of V^2 depends on the sample size n and often tends to σ^2 as $n \to \infty$. On the other hand, a scale mixture appears as a basic statistical

Y. Fujikoshi and V. V. Ulyanov, *Non-Asymptotic Analysis of Approximations for Multivariate Statistics*, JSS Research Series in Statistics,
https://doi.org/10.1007/978-981-13-2616-5_2

distribution. Then we are interested in the distance of the mixture from its parent distribution.

The approximation is based on knowledge of the underlying distribution G and information about some moments of the variable S. We tacitly assume that S is close to 1 in some sense. Not only do we propose approximation formulas for the cumulative distribution function (cdf) $F(x)$ and its density (pdf) $f(x)$ of the scale mixture X, but we also give the upper bounds of their possible errors evaluated in sup and L_1-norms and in terms of the moment $\mathrm{E}[(S^\rho - 1)^k]$, where ρ is a positive integer to be chosen suitably. Their possible errors evaluated in L_1-norms and an extension to multivariate scale mixtures shall be discussed in Chap. 3.

2.2 Error Bounds in Sup-Norm

If a random variable Z follows the distribution with the cdf $G(x)$ and the pdf $g(x)$, the cdf of the scale mixture $X = S^\delta Z$ is given by

$$F(x) \equiv \Pr\{X \le x\} = \Pr\{Z \le xS^{-\delta}\} = \mathrm{E}_s[\,G(xS^{-\delta})\,]. \tag{2.2}$$

Here δ is 1 or -1, which is decided by its usefulness. Assuming that the scale factor S is close to 1 in some sense, we consider approximating the cdf $F(x)$. Our interest also lies in evaluating possible errors of approximations. Throughout this chapter we assume without further notice that $G(x)$ is $k+1$ times continuously differentiable on its support $D = \{x \in R : g(x) > 0\}$ and that the scale factor S have moments of required order. For technical reasons, in the following, we use a random variable Y defined by

$$Y = S^\rho \quad \text{or} \quad S = Y^{1/\rho}, \tag{2.3}$$

where ρ is a suitably chosen positive integer, and mostly $\rho = 1$ or $\rho = 2$. For example, for t-distribution with n degrees of freedom, we choose $S = Y^{1/2}$ and $Y = \chi_n^2/n$.

For any $s > 0$, the conditional distribution of $X = S^\delta Z$ given $S = s$, or, $X = Y^{\delta/\rho} Z$ given $Y = y (= s^\rho)$ has a cdf that is expressible as $G(xy^{-\delta/\rho})$, and mathematical induction shows that its derivatives with respect to y can be put in the form

$$\frac{\partial^j G(xy^{-\delta/\rho})}{\partial y^j} = y^{-j} c_{\delta,j}(xy^{-\delta/\rho}) g(xy^{-\delta/\rho}), \quad j = 1, 2, \ldots, k, \tag{2.4}$$

where the $c_{\delta,j}$ are real functions obtained by

$$c_{\delta,j}(x) g(x) = \left. \frac{\partial^j G(xy^{-\delta/\rho})}{\partial y^j} \right|_{y=1}. \tag{2.5}$$

Let

$$\alpha_{\delta,j} \equiv \begin{cases} \max\{G(0), 1 - G(0)\}, & \text{for } j = 0, \\ (1/j!) \sup_x |c_{\delta,j}(x)|g(x), & \text{if } j \geq 1. \end{cases} \tag{2.6}$$

We can use

$$G_{\delta,k}(x, y) = G(x) + \sum_{j=1}^{k-1} \frac{1}{j!} c_{\delta,j}(x) g(x)(y-1)^j \tag{2.7}$$

as an approximation to $G(xy^{-\delta/\rho})$, which would in turn induce an approximation to $F(x)$ by means of

$$\begin{aligned} G_{\delta,k}(x) \equiv \mathrm{E}(G_{\delta,k}(x, Y)) &= G(x) + \sum_{j=1}^{k-1} \frac{1}{j!} c_{\delta,j}(x) g(x) \mathrm{E}[(Y-1)^j] \\ &= G(x) + \sum_{j=1}^{k-1} \frac{1}{j!} c_{\delta,j}(x) g(x) \mathrm{E}[(S^\rho - 1)^j]. \end{aligned} \tag{2.8}$$

In order to evaluate the approximation error of $G_{\delta,k}$ in (2.8), for $x \in D$ and $y > 0$, write

$$\Delta_{\delta,k}(x, y) \equiv G(xy^{-\delta/\rho}) - G_{\delta,k}(x, y). \tag{2.9}$$

By applying Taylor's theorem to $G(xy^{-\delta/\rho})$ as a function of y and using (2.7), the remainder term $\Delta_{\delta,k}(x, y)$ can be put in the form

$$\Delta_{\delta,k}(x, y) = \frac{1}{k!} c_{\delta,k}(u) g(u) y_0^{-k} (y-1)^k, \tag{2.10}$$

where $u = xy_0^{-\delta/\rho}$, and y_0 is a positive number lying between 1 and y. This implies the following lemma.

Lemma 2.1 *The residual*

$$\Delta_{\delta,k}(x, y) \equiv G(xy^{-\delta/\rho}) - G_{\delta,k}(x, y) \tag{2.11}$$

satisfies the inequality

$$\sup_x |\Delta_{\delta,k}(x, y)| \leq \alpha_{\delta,k}(y \vee y^{-1} - 1)^k, \quad \text{for each } y > 0. \tag{2.12}$$

This inequality cannot be improved in the sense that the constant $\alpha_{\delta,k}$ is the smallest possible value for which the inequality holds true.

The following theorem is an immediate consequence of the above lemma.

Theorem 2.1 *Let $X = S^{\delta} Z$ be a scale mixture. Suppose that S^{ρ} and $S^{-\rho}$ have k-moments. Then, we have*

$$\begin{aligned}|F(x) - G_{\delta,k}(x)| &\leq \alpha_{\delta,k} \mathrm{E}[(S^{\rho} \vee S^{-\rho} - 1)^k] \\ &\leq \alpha_{\delta,k} \{\mathrm{E}(|S^{\rho} - 1|^k) + \mathrm{E}(|S^{-\rho} - 1|^k)\}.\end{aligned}$$

Note that the approximation $G_{\delta,k}(x)$ depends on the moments of the S^{ρ} up to the $k - 1$ order, but the upper bound unpleasantly depends on the kth moment of $S^{-\rho}$ as well as S^{ρ}. In practical applications, we often face the situations where the moments of S^{ρ} are available, but $\mathrm{E}(S^{-\rho})$ is not, or even does not exist. It is important to find a bound of the form

$$|F(x) - G_{\delta,k}(x)| \leq b_{\delta,k} \mathrm{E}[|S^{\rho} - 1|^k].$$

In order to get this type of inequality we prepare two lemmas.

Lemma 2.2 *For all x and $0 < y < 1$,*

$$|\Delta_{\delta,k}(x, y)| \leq d_{\delta,k} \equiv \alpha_{\delta,0} + \alpha_{\delta,1} + \cdots + \alpha_{\delta,k-1}. \tag{2.13}$$

Proof The case $k = 1$ is clear. In general, we have for $k \geq 2$,

$$\begin{aligned}|\Delta_{\delta,k}(x, y)| &= \left| G(xy^{-\delta/\rho}) - G(x) - \sum_{j=1}^{k-1} \frac{1}{j!}(1 - y)^j c_{\delta,j}(x) g(x) \right| \\ &\leq |G(xy^{-\delta/\rho}) - G(x)| + \sum_{j=1}^{k-1} \frac{1}{j!} |c_{\delta,j}(x)| g(x) \leq d_{\delta,k}.\end{aligned}$$

□

Lemma 2.3 *Put*

$$\widetilde{\beta}_{\delta,k} = (\alpha_{\delta,k}^{1/k} + d_{\delta,k}^{1/k})^k. \tag{2.14}$$

Then the inequality

$$|G(xy^{-\delta/\rho}) - G_{\delta,k}(x, y)| \leq \widetilde{\beta}_{\delta,k} |y - 1|^k$$

holds for all real x and $y > 0$.

Proof Let c be any given constant between 0 and 1. Then, it follows from (2.10) that for $y \geq c$,

$$|\Delta_{\delta,k}(x, y)| \leq c^{-k} \alpha_{\delta,k} |y - 1|^k.$$

On the other hand, if $0 < y < c$, then we have

$$\begin{aligned}|\Delta_{\delta,k}(x,y)| &= |y-1|^{-k}|\Delta_{\delta,k}(x,y)||y-1|^k \\ &\le (1-c)^{-k} d_{\delta,k}|y-1|^k.\end{aligned}$$

Equating the right-hand sides of the above two inequalities, we find that $c = \{1 + (d_{\delta,k}/\alpha_{\delta,k})^{1/k}\}^{-1}$ is the best choice, and this completes the proof. □

From Lemma 2.3 we have the following theorem.

Theorem 2.2 *Let $X = S^\delta Z$ be a scale mixture. Suppose that S^ρ has kth moment. Then, we have*

$$|F(x) - G_{\delta,k}(x)| \le \widetilde{\beta}_{\delta,k} \mathrm{E}[|S^\rho - 1|^k], \tag{2.15}$$

where $\widetilde{\beta}_{\delta,k}$ is given by (2.14).

In order to look for an improvement on $\widetilde{\beta}_{\delta,k}$, let

$$J_{\delta,k}(x,y) \equiv \begin{cases} \dfrac{|G(xy^{-\delta/\rho}) - G_{\delta,k}(x,y)|}{|y-1|^k}, & \text{for } y \ne 1, \\[2ex] \dfrac{1}{k!}|c_{\delta,k}(x)|\ g(x), & \text{for } y = 1. \end{cases} \tag{2.16}$$

We define the positive constant $\beta_{\delta,k}$ by

$$\beta_{\delta,k} = \sup_{x \in D, y \le 1} J_{\delta,k}(x,y). \tag{2.17}$$

It follows from (2.10) that we have the inequalities $\alpha_{\delta,k} \le \beta_{\delta,k}$ and

$$\sup_x \left|\Delta_{\delta,k}(x,y)\right| \le \begin{cases} \alpha_{\delta,k} \left(y \vee y^{-1} - 1\right)^k, \\ \beta_{\delta,k} \left|y-1\right|^k. \end{cases} \tag{2.18}$$

For a set of positive numbers α, β, s (possibly random), a positive integer k and $\rho = 1$ or 2, such that $(s^\rho - 1)^k$ is nonnegative (a.s.), let the quantity $\Delta_k(\alpha, \beta;\ s, \rho)$ be defined by

$$\Delta_k(\alpha,\beta;\ s,\rho) \equiv \begin{cases} \beta \cdot (s^\rho - 1)^k, & \text{and} \\[1ex] \alpha \cdot (s^\rho - 1)^k, & \text{if } s \ge 1 \text{ (a.s.)}. \end{cases} \tag{2.19}$$

The following theorem is a direct consequence of the definition of quantities involved and the above observation.

Theorem 2.3 *Suppose that X is the scale mixture defined by* (2.1). *Then we have*

$$\sup_x |F(x) - G_{\delta,k}(x)| \le \mathrm{E}(\Delta_k(\alpha_{\delta,k}, \beta_{\delta,k};\ S, \rho)).$$

The case of $\delta = -1$ is noteworthy worth noticing in that a simple bound $\beta_{-1,k} = \alpha_0$ is available for all k, provided that an additional condition on G is satisfied, as described in Theorem 2.4. For the proof, see [3, 8].

Theorem 2.4 *Consider the case* $\delta = -1$ *under the same setup as in Theorem 2.3. Suppose further that G is analytic on D and that* $(-1)^{j+1}c_{-1,j}(x)$ *has the same sign as x for all x and j. Then we have*

$$|F(x) - G_{-1,k}(x)| \le \alpha_{-1,0}\mathrm{E}|S^\rho - 1|^k. \tag{2.20}$$

The inequality cannot be improved.

For an important application of Theorem 2.4, see Theorem 2.5.

2.3 Special Cases

2.3.1 Scale-Mixed Normal

By taking $\rho = 2$, the previous results applied to the normal case are summarized in the following theorem. We write Φ to denote the distribution function of the standard normal distribution $N(0, 1)$ and Φ its density. For $j = 1, 2, \ldots$, let $Q_{\delta,2j-1}(x)$ be the polynomials defined by

$$Q_{1,2j-1}(x) = H_{2j-1}(x), \tag{2.21}$$

$$Q_{-1,2j-1}(x) = K_{2j-1}(x) \equiv \sum_{m=0}^{j-1}(2m-1)!!\binom{j-1}{m}x^{2(j-m)-1}, \tag{2.22}$$

where we use the notation $(2m-1)!! = (2m-1)\cdot(2m-3)\cdots 3\cdot 1$ with the convention $(-1)!! = 1$, and $H_j(x)$ is the Hermite polynomial:

$$H_j(x)\Phi(x) = (-1)^j\frac{d^j}{dx^j}\Phi(x).$$

The polynomials H and K for $1 \le j \le 3$ are given as follows:

$$H_1(x) = x,\ H_3(x) = x^3 - 3x,\ H_5(x) = x^5 - 10x^3 + 5x,$$
$$K_1(x) = x,\ K_3(x) = x^3 + 3x,\ K_5(x) = x^5 + 2x^3 + 3x.$$

Now we have

Theorem 2.5 *The cdf $F(x)$ of the scale mixture $X = S^\delta Z$ of the* N(0, 1) *is approximated by*

$$\Phi_{\delta,k}(x) \equiv \Phi(x) - \sum_{j=1}^{k-1} \frac{\delta^j}{2^j j!} Q_{\delta,2j-1}(x) \mathrm{E}[(S^2-1)^j]\varphi(x). \tag{2.23}$$

The error estimate corresponding to (2.24) *is*

$$\sup_x |F(x) - \Phi_{\delta,k}(x)| \le \mathrm{E}(\Delta_k(\alpha_{\delta,k}, \beta_{\delta,k};\ s, 2)). \tag{2.24}$$

Here, $\beta_{-1,k} = 1/2$, and $\beta_{1,k} = 1/2$ for $k \le 6$.

Proof We consider the transform $S \to Y = S^2$. Let the real function $c_{\delta,j}(x)$ be defined by

$$c_{\delta,j}(x)\varphi(x) = \frac{\partial^j}{\partial y^j} \Phi(xy^{-\delta/2})\Big|_{y=1}. \tag{2.25}$$

Then writing

$$Q_{\delta,2j-1}(x) = -(2\delta)^j c_{\delta,j}(x),$$

we obtain (2.22). The defining formula (2.22) for $K_{2j-1}(x)$ follows from

$$\frac{\partial^j}{\partial y^j}\Phi(xy^{1/2}) = \frac{(-1)^{j-1}}{2^j} \sum_{m=0}^{j-1} (2m-1)!! \binom{j-1}{m} x^{2(j-m)-1} y^{-\frac{1}{2}-m} \varphi(xy^{1/2}),$$

which will be proved by mathematical induction.

We can use Theorem 2.5 to obtain $\beta_{-1,k} = 1/2$. We have only to note that

$$(-1)^{j+1} c_{-1,j}(x) = \frac{1}{2^j} K_{2j-1}(x) \begin{cases} \ge 0 & \text{if } x \ge 0, \\ < 0 & \text{if } x < 0. \end{cases}$$

Here, $\beta_{1,k} = 1/2 (k \le 6)$ was obtained by evaluating $J_{1,k}(y)$ of (2.16) numerically. □

Numerical values of $\alpha_{\delta,k}$ are given in Table 2.1 for $k \le 6$. We can take $\beta_{\delta,k} = 1/2$ for $\delta = 1$, $k \le 6$ and $\beta_{-1,k} = 1/2$ for all $k = 1, 2, \ldots$.

Example 2.1 *(t-Distribution)* Let Z and χ^2_n be independent random variables following the standard normal N(0, 1) and the chi-square distribution with n degrees of freedom, respectively, and put $S = \sqrt{\chi^2_n/n}$. Then the scale mixture $S^{-1}Z$ follows the t-distribution with n degrees of freedom, which is approximated by (2.23) with $\delta = -1$. For $j \ge 1$ we have

Table 2.1 Numerical values of $\alpha_{\delta,k}$ for N(0, 1)

k	1	2	3	4	5	6
$\alpha_{1,k}$	0.1210	0.0688	0.0481	0.0370	0.0300	0.0252
$\alpha_{-1,k}$	0.1210	0.0791	0.0608	0.0501	0.0431	0.0380

$$\mathrm{E}(S^{2j}) = \frac{n(n+2)\cdots(n+2j-2)}{n^j}.$$

Taking $k=4$, $q=4$, and $\delta=-1$, and noting that

$\alpha_{-1,4} = 0.0501$, $\beta_{-1,4} = 0.5$
$K_3(x) = x^3 + x, \quad K_5(x) = x^5 + 2x^3 + 3x,$
$\mathrm{E}[(S^2-1)] = 0, \quad \mathrm{E}[(S^2-1)^2] = 2/n,$
$\mathrm{E}[(S^2-1)^3] = 8/n^2$, and $\mathrm{E}[(S^2-1)^4] = 12(n+4)/n^3$,

we obtain

$$\Phi_{-1,4}(x) = \Phi(x) - \left(\frac{x^3+x}{4n} - \frac{x^5+2x^3+3x}{6n^2}\right)\Phi(x)$$

and

$$\sup|F_n(x) - \Phi_{-1,4}(x)| \le \frac{6(n+4)}{n^3}.$$

There are considerably many estimators which are expressed as scale-mixed normal $X = SZ$ with $S \ge 1$. For example, these appear as estimators (see, e.g., [3]) in growth curve model, profile analysis, seemingly unrelated model. Let $X = SZ$ be a scale mixture of the standard normal distribution with the scale factor $S \ge 1$. For such scale mixture, by using Theorem 2.3 in the case of $\delta = 1$ we obtain an asymptotic expansion approximation and its error bound. In the following we derive them by a different approach. First, note that the X in (2.1) can be written as

$$X = Z - U, \tag{2.26}$$

where $U = \sqrt{S^2-1}V$, $V \sim \mathrm{N}(0,1)$, and V is independent of Z and S. Therefore, we have

$$F(x) \equiv \Pr\{SZ \le x\} = \Pr\{Z - U \le x\} = \mathrm{E}_{S,V}[\Phi(x+U)].$$

Consider the Taylor expansion given by

$$\Phi(x+u) = \sum_{j=0}^{2k-1} \frac{1}{j!}\Phi^{(j)}(x)u^j + R_{2k},$$

where $R_{2k} = \left((2k)!\right)^{-1}\Phi^{(2k)}(x+\theta u)u^{2k}$ and $0 < \theta < 1$. We can see that

$$\mathrm{E}(V^j) = \begin{cases} \dfrac{(2m)!}{2^m m!}, & \text{if } j = 2m \text{ (even)}, \\ 0, & \text{if } j = 2m-1 \text{ (odd)}. \end{cases}$$

The result is obtained by checking that

$$\mathrm{E}_{s,V}\left[\sum_{j=0}^{2k-1}\frac{1}{j!}\Phi^{(j)}(x)U^j\right] = \Phi_{1,k}(x),$$
$$|R_{2k}| \le \alpha_{1,k}\mathrm{E}[(S^2-1)^k].$$

It was noted by [1] that an estimator in the growth curve model has a structure as in (2.26) and gave an error bound.

2.3.2 Scale-Mixed Gamma

The class $G(\sigma,\lambda)$ of gamma distributions is a two-parameter family consisting of all distributions with the density

$$g(x;\sigma,\lambda) = \frac{1}{\sigma\Gamma(\lambda)}\left(\frac{x}{\sigma}\right)^{\lambda-1}\mathrm{e}^{-x/\sigma}, \quad x > 0, \quad \sigma, \ \lambda > 0. \tag{2.27}$$

We consider the case that the underlying distribution is $G(1,\lambda)$ with the pdf $g(x;\lambda) \equiv g(x;1,\lambda)$. Then, for the approximation of the cdf $F(x;\lambda)$ of the scale mixture $X = S^\delta Z$, we take $\rho = 1$. The approximating function will be given in terms of the Laguerre polynomial defined by $L_{(\lambda,0)}(x) = 1$ and

$$\begin{aligned} L_{(\lambda,p)}(x) &\equiv (-1)^p x^{-\lambda}\mathrm{e}^x\frac{d^p}{dx^p}\left(x^{p+\lambda}\mathrm{e}^{-x}\right) \\ &= \sum_{\ell=0}^{p}(-1)^\ell(p+\lambda)^{(\ell)}\binom{p}{\ell}x^{p-\ell}. \end{aligned} \tag{2.28}$$

Setting

$$L_{1,j}(x;\lambda) = xL_{(\lambda,j-1)}(x) \quad \text{and} \quad L_{-1,j}(x;\lambda) = xL_{(\lambda-j,j-1)}(x),$$

the approximating function for $G(xy^{-\delta};\lambda)$ corresponding to (2.7) is given by

$$G_{\delta,k}(x,y;\lambda) \equiv G(x;\lambda) - \sum_{j=1}^{k-1} \frac{\delta^j}{j!} L_{\delta,j}(x;\lambda) g(x;\lambda)(y-1)^j. \tag{2.29}$$

Let $\alpha_{\delta,k}(\lambda)$ and $\beta_{\delta,k}(\lambda)$ be positive numbers defined by

$$\alpha_{\delta,k}(\lambda) = \sup_x \left| \frac{1}{k!} L_{\delta,k}(x;\lambda) g(x;\lambda) \right|,$$
$$\beta_{\delta,k}(\lambda) = \sup_{x,0<y<1} \left| \frac{G(xy^{-\delta};\lambda) - G_{\delta,k}(x,y;\lambda)}{(y-1)^k} \right|. \tag{2.30}$$

Then, from (2.4) and (2.7) we have the following theorem by setting $\rho = 1$.

Theorem 2.6 *The cdf $F(x;\lambda)$ of the scale mixture $X = S^\delta Z$ of the gamma distribution $G(1;\lambda)$ is approximated by*

$$G_{\delta,k}(x;\lambda) \equiv G(x;\lambda) - \sum_{j=1}^{k-1} \frac{\delta^j}{j!} L_{\delta,j}(x;\lambda) \mathrm{E}[(S-1)^j] g(x;\lambda). \tag{2.31}$$

The error estimate corresponding to (2.3) *is*

$$\sup_x |F(x;\lambda) - G_{\delta,k}(x;\lambda)| \le \mathrm{E}(\Delta_k(\alpha_{\delta,k}(\lambda), \beta_{\delta,k}(\lambda);\ S, 1)). \tag{2.32}$$

Values of $\beta_{\delta,k}(\lambda)$ can be obtained by maximizing the ratio $J_{\delta,k}(x,y;\lambda)$ of (2.16) numerically. Note, however, that $\beta_{-1,k}(0.5) = \beta_{-1,k}(1.0) = 1$ follows from Theorem 2.4 and that $\beta_{\delta,1}(\lambda) = 1$ is a simple consequence of the fact that $G_{\delta,1}(x;\lambda)$ is a cdf for all $\lambda > 0$. The chi-square distribution with q degrees of freedom is a special case $G(2; q/2)$ of the gamma distribution, with the density

$$\begin{aligned} g_q(x) &= \frac{1}{2} g(x/2;\ q/2) \\ &= \frac{1}{2\Gamma(q/2)} \left(\frac{x}{2}\right)^{q/2-1} \mathrm{e}^{-x/2}, \quad x > 0,\ q = 1, 2, \ldots. \end{aligned} \tag{2.33}$$

Therefore, an approximating function for the cdf $F_{\delta,q}(x)$ of scale mixture $X = S^\delta Z$ of χ^2_q is derived from (2.29). In fact, given the condition $S = s$, the conditional cdf of X is $G(xs^{-\delta}/2,\ q/2)$, and an approximation is given by

$$G_{\delta,k;q}(x) \equiv G(x/2;\ q/2) - \sum_{j=1}^{k-1} \frac{\delta^j}{j!} 2\, L_{\delta,j}\Big(\frac{x}{2};\ \frac{q}{2}\Big) \cdot \frac{1}{2} g\Big(\frac{x}{2};\ \frac{q}{2}\Big) \cdot \mathrm{E}[(S-1)^j]$$

$$= G_q(x) - \sum_{j=1}^{k-1} \frac{\delta^j}{2^{j-1} j!} U_{\delta,j;q}(x)\, \mathrm{E}[(S-1)^j]\, g_q(x), \qquad (2.34)$$

where $U_{\delta,j;q}(x)$ is the polynomial defined by

$$U_{\delta,j;q}(x) \equiv L_{\delta,j}\big(x/2;\ q/2\big), \qquad (2.35)$$

and approximation errors are bounded by

$$\begin{aligned} \sup_x |F_{\delta,q}(x) - G_{\delta,k;\,q}(x)| &= \sup_x |F(x/2;\ q/2) - G_{\delta,k}(x/2;\ q/2)| \\ &\le \mathrm{E}(\Delta_k(\alpha_{\delta,k;q}, \beta_{\delta,k;q};\ S, 1)) \\ &\equiv \mathrm{E}(\Delta_k(\alpha_{\delta,k}(q/2), \beta_{\delta,k}(q/2);\ S, 1)). \end{aligned} \qquad (2.36)$$

This means that numerical values of α and β are available from the ones for $G(1, \lambda)$, by reading $\alpha_{\delta,k;q}$ for $\alpha_{\delta,k}(q/2)$ and $\beta_{\delta,k;q}$ for $\beta_{\delta,k}(q/2)$, respectively. For example, if $k = 4$, the bound becomes

$$\sup_x |F_{-1,4}(x) - G_{-1,4;\,4}(x)| \le \beta_{-1,4;\,4} \cdot \mathrm{E}[(S-1)^4] = \mathrm{E}[(S-1)^4].$$

If $S > 1$ with probability 1, then $\beta_{-1,4;\,4}$ can be replaced by $\alpha_{-1,4;\,4} = 0.4016$.

Example 2.2 *(F-distribution)* Let χ^2_q and χ^2_n be mutually independent chi-squared variables with q and n degrees of freedom, respectively. Instead of the F variable $\chi^2_q/q/(\chi^2_n/n)$, we consider the scale mixture $T = \chi^2_q/(\chi^2_n/n)$ of the chi-square distribution, which appears in Chap. 3. Taking, for instance, $k = 4$, $q = 4$, and $\delta = -1$, and noting that

$$\begin{aligned} &G_4(x) = 1 - (x/2+1)\mathrm{e}^{-x/2}, \quad g_4(x) = (x/4)\mathrm{e}^{-x/2}, \quad \beta_{-1,4}(2) = 1.0, \\ &U_{-1,1;\,4}(x) = x, \quad U_{-1,2;\,4}(x) = x(x-2), \quad U_{-1,3;\,4}(x) = x(x-2)(x+2), \\ &\mathrm{E}[(S-1)] = 0, \quad \mathrm{E}[(S-1)^2] = 2/n, \\ &\mathrm{E}[(S-1)^3] = 8/n^2, \quad \mathrm{E}[(S-1)^4] = 12(n+4)/n^3, \end{aligned}$$

we obtain from (2.34) and (2.36) the fact that the cdf of T is approximated by

$$G_{-1,4;\,4}(x) = G_4(x) - \frac{x(x-2)}{n}\left(\frac{1}{2} - \frac{1}{3n}(x+2)\right) g_4(x)$$

with the approximation error bounded by

$$\Delta_4(n) = \sup \big| \Pr\{T \le x\} - G_{-1,4;\,4}(x) \big| \le \frac{12(n+4)}{n^3}.$$

2.3.3 Scale-Mixed F

We present some basic results on asymptotic expansions and their error bounds for a scale mixture of an F-distribution defined by

$$X_{q,n} = S_n^{\delta} Z_{q,n}, \tag{2.37}$$

where $Z_{q,n}/q$ is a random variable with an F-distribution with (q, n) degrees of freedom, S_n is a positive random variable, and S_n and $Z_{q,n}$ are independent. Here $\delta = \pm 1$ is a constant used to distinguish two types of scale mixtures: $X_{q,n} = S_n Z_{q,n}$ and $X_{q,n} = S_n^{-1} Z_{q,n}$. A scale mixture of an F-distribution appears, for example, in profile analysis (see, [3, 9]). In fact, consider profile analysis of k p-variate normal populations based on N samples. Then, it is known that a simultaneous confidence interval for differences in the levels of k profiles is based on a statistic T, which is expressed as

$$T = \left(1 + \frac{\chi_{p-1}^2}{\chi_{n+1}^2}\right) \frac{\chi_q^2}{\chi_n^2/n}, \tag{2.38}$$

where $q = k - 1$ and $n = N - k - p + 1$. Here χ_q^2, χ_n^2, χ_{p-1}^2, and χ_{n+1}^2 are independent. Then, we can express T as a sale mixture of F-distribution in two ways with

$$\delta = 1, \quad S_n = 1 + \frac{\chi_{p-1}^2}{\chi_{n+1}^2}, \quad Z_{q,n} = \frac{\chi_q^2}{\chi_n^2/n}, \tag{2.39}$$

and

$$\delta = -1, \quad S_n = \frac{\chi_{n+1}^2}{\chi_{n+1}^2 + \chi_{p-1}^2}, \quad Z_{q,n} = \frac{\chi_q^2}{\chi_n^2/n}. \tag{2.40}$$

If we use (2.39) with $\delta = 1$, $S_n > 1$, and if we use (2.40) with $\delta = -1$, $S_n < 1$.

We present asymptotic expansions of $X_{q,n}$ in (2.37) and their error bounds. The results are applied to the distribution of T in (2.38). Let $F_q(x; n)$ and $f_q(x; n)$ be the cdf and the pdf of $Z_{q,n}$. Then, since $Z_{q,n}/q$ is an F-distribution with (q, n) degrees of freedom, with $\delta = \pm 1$, the density is given by

$$f_q(x;\, n) = B_0(q, n) \frac{1}{n} \left(\frac{x}{n}\right)^{q/2-1} \left(1 + \frac{x}{n}\right)^{-(q+n)/2},$$

where

$$B_0(q, n) \equiv \frac{\Gamma((q+n)/2)}{\Gamma(q/2)\,\Gamma(n/2)}.$$

This means that $Z_{q,n}/q$ follows an F-distribution $F(q, n)$ with (q, n) degrees of freedom. Using (2.8), we have an approximation of the cdf $G_q(x;\, n)$ of the distribution of $X_{q,n}$ for large n and its error bound. For detail, see [2, 3].

2.4 Error Bounds Evaluated in L_1-Norm

In this section, we study the asymptotic approximation of the pdf of the distribution of the scale mixture $X = S^\delta Z$ and evaluate its error in the L_1-norm, which will make it possible to approximate the probability $\Pr\{X \in A\}$ for an arbitrary Borel set A. An extension to multivariate cases shall be discussed in Chap. 3. Suppose that the underlying distribution function G of X has the density $g(x)$, and let the random variable Y be defined by $Y = S^\rho$. Then the conditional density function of X given the condition $S = s$ or $Y = y(= s^\rho)$ is

$$g(x \mid y) = y^{-\delta/\rho} g(xy^{-\delta/\rho}).$$

Therefore, the density $f(x)$ of X is given by

$$f(x) = \mathrm{E}_Y[g(x \mid Y)] = \mathrm{E}_Y[Y^{-\delta/\rho} g(xY^{-\delta/\rho})]. \tag{2.41}$$

Assuming that the density g is k-times continuously differentiable on its support $D \equiv \{x \in R : g(x) > 0\}$, we introduce a set of functions $b_{\delta,j}(x)$, $j = 0, 1, \ldots, k-1$ by $b_{\delta,0}(x) = 1$ and

$$\frac{\partial^j}{\partial y^j}\left(y^{-\delta/\rho} g(xy^{-\delta/\rho})\right) = y^{-j} b_{\delta,j}(xy^{-\delta/\rho}) \cdot y^{-\delta/\rho} g(xy^{-\delta/\rho}),$$

or

$$b_{\delta,j}(x)g(x) = \frac{\partial^j}{\partial y^j}\left(y^{-\delta/\rho} g(xy^{-\delta/\rho})\right)\Big|_{y=1}. \tag{2.42}$$

For each $x \in R$, we take the first k terms $g_{\delta,k}(x, y)$ of the Taylor expansion of $g(x \mid y)$ around $y = 1$ as its approximation:

$$g_{\delta,k}(x, y) = \sum_{j=0}^{k-1} \frac{1}{j!} b_{\delta,j}(x)g(x)(y-1)^j. \tag{2.43}$$

This suggests that the function

$$\begin{aligned} g_{\delta,k}(x) &\equiv \mathrm{E}[g_{\delta,k}(x, Y)] \\ &= \sum_{j=0}^{k-1} \frac{1}{j!} b_{\delta,j}(x)\mathrm{E}[(S^\rho - 1)^j]g(x) \end{aligned} \tag{2.44}$$

provides a reasonable approximation of the density $f(x)$. Comparing both sides of (2.8) and (2.44), we obtain

$$b_{\delta,j}(x)g(x) = \frac{d}{dx}\big(c_{\delta,j}(x)g(x)\big). \tag{2.45}$$

In fact, the approximating function $g_{\delta,k}(x)$ is obtained by differentiating $G_{\delta,k}(x)$ of (2.8). Note that $g_{\delta,k}(x)$ is not necessarily a pdf, but it satisfies the relation

$$\int_R g_{\delta,k}(x)dx = G_{\delta,k}(\infty) - G_{\delta,k}(-\infty) = 1. \tag{2.46}$$

Our error bounds depend on

$$\xi_{\delta,j} = \frac{1}{j!}\left\|b_{\delta,j}(x)g(x)\right\|_1, \ (j = 0, 1, \ldots, k) \tag{2.47}$$

and more precisely they are expressed in terms of

$$\eta_{\delta,k} = \left\{\xi_{\delta,k}^{1/k} + \left(2 + \sum_{j=1}^{k-1}\xi_{\delta,j}\right)^{1/k}\right\}^k. \tag{2.48}$$

Here for any integrable function $h(\boldsymbol{x}) : \mathbb{R}^p \to \mathbb{R}^1$, we define its L_1-norm by

$$\|h(\boldsymbol{x})\|_{1;p} = \int_{\mathbb{R}^p} |h(\boldsymbol{x})|d\boldsymbol{x},$$

and in particular, we write $\|\cdot\|_1 = \|\cdot\|_{1;1}$. Our error estimate is based on an evaluation of the quantity

$$\Xi_{\delta,k}(x, y) \equiv y^{-\delta/\rho}g(xy^{-\delta/\rho}) - g_{\delta,k}(x, y) \ \left(= \frac{\partial}{\partial x}\Delta_{\delta,k}(x, y)\right), \tag{2.49}$$

where $\Delta_{\delta,k}(x, y)$ is defined by (2.9).

Theorem 2.7 *Let $X = S^\delta Z$ be a scale mixture. Suppose that S^ρ has kth moment. Then, for any $k \geq 1$ it holds that*

$$\left\|f(x) - g_{\delta,k}(x)\right\|_1 \leq \frac{1}{2}\eta_{\delta,k}\mathrm{E}\left[|Y - 1|^k\right], \tag{2.50}$$

where $\eta_{\delta,k}$ is given by (2.48).

Proof We use a Taylor formula (see, e.g., [6, p. 257]) for a function h with $k(\geq 1)$ continuous derivatives

$$h(y) = h(1) + \sum_{j=1}^{k-1} \frac{1}{j!} h^{(j)}(1)(y-1)^j + \frac{(y-1)^k}{(k-1)!} \mathrm{E}\left[(1-\tau)^{k-1} h^{(k)}(1+\tau(y-1))\right], \tag{2.51}$$

where τ is a random variable with uniform distribution (0, 1). Using (2.42), (2.49), and (2.51) we can write also for $k \geq 1$

$$\Xi_{\delta,k}(x,y) = \frac{(y-1)^k}{(k-1)!} \mathrm{E}\left[(1-\tau)^{k-1} (1+\tau(y-1))^{-k-\delta\rho} \times b_{\delta,k}\left(x(1+\tau(y-1))^{-\delta\rho}\right) g\left(x(1+\tau(y-1))^{-\delta\rho}\right)\right]. \tag{2.52}$$

The idea of our proof is to use (2.49) or (2.52) depending on whether y is far from 1 or close to it. Let

$$\varphi = (\xi_{\delta,k}/\eta_{\delta,k})^{1/k}.$$

Note that $\varphi : 0 < \varphi < 1$. If $y : 0 < y < \varphi$, then it follows from (2.9) that

$$\left\| \Delta_{\delta,k}(x,y) \right\|_1 \leq \left(1 + \sum_{j=0}^{k-1} \xi_{\delta,j} \right) \frac{(1-y)^k}{(1-\varphi)^k} = \eta_{\delta,k} |y-1|^k. \tag{2.53}$$

If $y \geq \varphi$, then for any $\tau \in [0,1]$ we have $1 + \tau(y-1) \geq \varphi$. Therefore it follows from (2.52) and Fubini theorem that

$$\left\| \Xi_{\delta,k}(x,y) \right\|_1 \leq \xi_{\delta,k} \frac{|y-1|^k}{\varphi^k} = \eta_{\delta,k} |y-1|^k. \tag{2.54}$$

Combining (2.53) and (2.54) we get (2.50). □

In the last part in the proof of Theorem 2.7, we use the following result. Let X be a random variable with the pdf f. Let g be an integrable real function such tat $\int g(x)dx = 1$, but that is not necessarily a pdf (i.e., g(x) may take negative values). Then, it holds (see Problem 13.4 in [3]) that

$$\left| \Pr\{X \in A\} - \int_A g(x)dx \right| \leq \frac{1}{2} \int_{-\infty}^{\infty} |f(x) - g(x)|dx \tag{2.55}$$

for any Borel subset $A \subset \mathbb{R}$.

In order to get an improved error estimate, writing

$$\Xi_{\delta,k}(y) \equiv \int_R |g(x|\, y) - g_{\delta,k}(x, y)|dx \quad \left(= \int_R |\Xi_{\delta,k}(x,\ y)|dx\right), \tag{2.56}$$

let

$$\sigma_{\delta,k} = \frac{1}{k!}\left\|b_{\delta,k}(x)g(x)\right\|_1 \text{ and } \tau_{\delta,k} = \sup_{0<y<1} \frac{\Xi_{\delta,k}(y)}{|y-1|^k}. \tag{2.57}$$

These quantities correspond to the α and β described in Sect. 2.1.

We prove the following:

Lemma 2.4 *It holds that for $k \geq 1$ and $y > 0$*

(1) $\Xi_{\delta,k}(y) \leq \sigma_{\delta,k}\ (y \vee y^{-1} - 1)^k$,
(2) $\Xi_{\delta,k}(y) \leq \tau_{\delta,k}\ |y - 1|^k$.

Proof To prove (1), we write the residual $\Xi_{\delta,k}(x, y)$ of the Taylor expansion in integral form:

$$\begin{aligned}\Xi_{\delta,k}(x, y) &= y^{-\delta/\rho}g(xy^{-\delta/\rho}) - \sum_{j=0}^{k-1}\frac{1}{j!}b_{\delta,j}(x)g(x)(y-1)^j \\ &= \frac{1}{(k-1)!}\int_1^y z^{-k}b_{\delta,k}(xz^{-\delta/\rho})\cdot z^{-\delta/\rho}g(xz^{-\delta/\rho})(y-z)^{k-1}dz.\end{aligned}$$

Then we can use Fubini theorem together with the definition of $\sigma_{\delta,k}$ to obtain

$$\begin{aligned}\Xi_{\delta,k}(y) = \int_R |\Xi_{\delta,k}(x, y)|dy &\leq \frac{k!}{(k-1)!}\sigma_{\delta,k}\left|\int_1^y z^{-k}(y-z)^{k-1}dz\right| \\ &\leq \frac{k!}{(k-1)!}\sigma_{\delta,k}(y^{-k} \vee 1)\left|\int_1^y (y-z)^{k-1}dz\right| \\ &= \sigma_{\delta,k}\ (y \vee y^{-1} - 1)^k,\end{aligned}$$

as was to be proved. Assertion (2) is clear from the definition of τ. □

The following theorem is a direct consequence of the definition of quantities involved and Lemma 2.4.

Theorem 2.8 *Suppose that $X = S^\delta Z$ is a scale mixture of Z and has the pdf $f(x)$. Then we have for any Borel set $A \subset \mathbb{R}$,*

$$\begin{aligned}\left|\Pr(X \in A) - \int_A g_{\delta,k}(x)dx\right| &\leq \frac{1}{2}\left\|f(x) - g_{\delta,k}(x)\right\|_1 \\ &\leq \frac{1}{2}\mathrm{E}\left[\Delta_k(\sigma_{\delta,k}, \tau_{\delta,k};\ S, \rho)\right],\end{aligned} \tag{2.58}$$

where $\Delta_k(\sigma, \tau;\ s, \rho)$ is defined by (2.19).

References

1. Fujikoshi, Y. (1985). An error bound for an asymptotic expansion of the distribution function of an estimate in a multivariate linear model. *The Annals of Statistics*, *13*, 827–831.
2. Fujikoshi, Y., & Shimizu, R. (2018). *Asymptotic expansions for scale mixtures of F-distribution and their error bounds*. Hiroshima Statistical Research Group, TR; 17-2.
3. Fujikoshi, Y., Ulyanov, V. V., & Shimizu, R. (2010). *Multivariate analysis: High-dimensional and large-sample approximations*. Hoboken: Wiley.
4. Gleser, L. J., & Olkin, I. (1972). *Estimation for regression model with an unknown covariance matrix* (pp. 541–568). Proc. Sixth Berkeley Symposium on Mathematical Statistics and Probability I: University of California, Berkeley, CA.
5. Kariya, T., & Maekawa, K. (1982). A method for approximations to the pdf's and cdf's of GLSE's and its application to the seemingly unrelated regression model. *Annals of the Institute of Statistical Mathematics A*, *34*, 281–297.
6. Khuri, A. I. (2003). *Advanced calculus with applications in statistics* (2nd ed.). Hoboken: Wiley.
7. Shimizu, R. (1987). Error bounds for asymptotic expansion of the scale mixtures of the normal distribution. *Annals of the Institute of Statistical Mathematics*, *39*, 611–622.
8. Shimizu, R., & Fujikoshi, Y. (1997). Sharp error bounds for asymptotic expansions of the distribution functions of scale mixtures. *Annals of the Institute of Statistical Mathematics*, *49*, 285–297.
9. Srivastava, M. S. (2002). *Methods of multivariate statistics*. Hobohen: Wiley.

Chapter 3
MANOVA Test Statistics

Abstract The main purpose of this chapter is to give a method for obtaining error bounds for asymptotic expansions of the null distributions of Hotelling's T^2 (or Lawley–Hotelling criterion, T_{LH}), the likelihood-ratio criterion T_{LR} and the Bartlett–Nanda–Pillai criterion T_{BNP} in the MANOVA model when the sample size is large. The results for T_{LH} and T_{LR} are obtained by expressing these statistics in terms of a multivariate scale mixture, and using error bounds evaluated in L_1-norm. The error bound is given for the limiting distribution of T_{BNP} by using a relationship between T_{BNP} and T_{LH}. Further, we give error bounds for these criteria when the sample size and the dimension are large.

3.1 Introduction

We consider three test criteria: the likelihood-ratio criterion T_{LR}, the Lawley–Hotelling criterion T_{LR}, and the Bartlett–Nanda–Pillai criterion T_{BNP} used for testing in MANOVA and the multivariate linear model. Without loss of generality, we may represent the null distributions of these criteria in the following form:

$$\begin{aligned} T_{LR} &= -n\log(|\mathbf{S}_e|/|\mathbf{S}_e+\mathbf{S}_h)|), \\ T_{LH} &= n\mathrm{tr}\mathbf{S}_h\mathbf{S}_e^{-1}, \\ T_{BNP} &= n\mathrm{tr}\mathbf{S}_h(\mathbf{S}_h+\mathbf{S}_e)^{-1}. \end{aligned} \tag{3.1}$$

Here, $\mathbf{S}_h$ and $\mathbf{S}_e$ are the matrices of sums of squares and products due to hypothesis and error, respectively. Under hypothesis, $\mathbf{S}_h$ and $\mathbf{S}_e$ are independently distributed as Wishart distributions $\mathrm{W}_p(q, \mathbf{I}_p)$ and $\mathrm{W}_p(n, \mathbf{I}_p)$, respectively, where n is the number depending the sample size, p is the dimension of the observations or response variables, and q is the number of treatments or explanatory variables. It is assumed that $n \geq p$.

Under a large sample asymptotic framework, $n \to \infty$ and fixed p and q, it is known (see, e.g., [7]) that

Y. Fujikoshi and V. V. Ulyanov, *Non-Asymptotic Analysis of Approximations for Multivariate Statistics*, JSS Research Series in Statistics,
https://doi.org/10.1007/978-981-13-2616-5_3

$$\Pr(T \le x) = G_r(x) + R_1, \tag{3.2}$$

and

$$\Pr(T \le x) = G_r(x) + \frac{r}{4n}\{a_0 G_r(x) + a_1 G_{r+2}(x) + a_2 G_{r+4}(x)\} + R_2, \tag{3.3}$$

where G_r is the distribution function of a χ_r^2-variate, and $r = pq$. The constants a_i are given as follows:

$$\begin{aligned}
T_{LR} &: a_0 = q - p - 1, \quad a_1 = -(q - p - 1), \quad a_2 = 0. \\
T_{LH} &: a_0 = q - p - 1, \quad a_1 = -2q, \quad a_2 = q + p + 1, \\
T_{BNP} &: a_0 = q - p - 1, \quad a_1 = -2q, \quad a_2 = q + p + 1.
\end{aligned}$$

These expansions were first obtained by expanding their characteristic functions and inverting them formally. After that, a rigorous validity was established by [1]. The validity means that the remainder terms R_1 and R_2 satisfy

$$R_1 = \mathrm{O}(n^{-1}), \quad R_2 = \mathrm{O}(n^{-2}). \tag{3.4}$$

In order to overcome a weak point of error estimates discussed in Chap. 1, we present a method of obtaining error bounds $B_{i,p}(q, n)$ $(i = 1, 2)$:

$$|R_i| \le B_{i,p}(q, n) \tag{3.5}$$

such that (1) $B_{i,p}(q, n) = \mathrm{O}(n^{-i})$, and (2) $B_{i,p}(q, n)$ is explicit and computable. In this chapter, we present an outline for obtaining error bounds with properties (1) and (2) for T_{LH}, T_{LR}, and T_{BNP}. For simplicity, the case $i = 1$ is considered in detail. The results for T_{LH} and T_{LR} were obtained by [2, 3]. An error bound for T_{BNP} was obtained by [4], by using a relation of T_{LH}.

In our derivation of T_{LH} and T_{LR} we use the fact that these statistics are expressed as a function of multivariate scale mixture given by

$$\mathbf{X} = \mathbf{S}^{\delta}\mathbf{Z} = \mathrm{diag}(S_1^{\delta}, \ldots, S_p^{\delta})\mathbf{Z}, \tag{3.6}$$

where δ is 1 or -1, the S_i's are positive random variables, and $\mathbf{S} = \mathrm{diag}(S_1, \ldots, S_p)$ is independent of $\mathbf{Z} = (Z_1, \ldots, Z_p)'$. We also use $Y_i = S_i^{\rho}, i = 1, \ldots, p$, where ρ is a suitably chosen positive integer, and mostly $\rho = 1$ or $\rho = 2$. In our application,

$$Z_1, \ldots, Z_q \sim i.i.d.\ \chi_q^2, \quad \delta = -1, \quad \rho = 1. \tag{3.7}$$

We present a theory of approximation of the distribution of $\boldsymbol{X}$ and the error bounds thereof. An error bound for T_{BNP} will be found by using a close relationship between T_{BNP} and T_{LH}, and using an error bound for T_{LH}.

In the last section, we treat error bounds for high-dimensional approximations of the three test statistics when $p/n \to c \in (0, 1)$.

3.2 Multivariate Scale Mixtures for T_{LH} and T_{LR}

First we shall see that T_{LH} and T_{LR} can be expressed as a function $T = T(X)$ of multivariate scale mixture

$$X = (X_1, \ldots, X_p)', \quad X_i = S_i^{-1} Z_i, \quad i = 1, \ldots, p, \tag{3.8}$$

where $Z_1, \ldots, Z_p \sim i.i.d.\ \chi_q^2$, $S_i > 0\ (i = 1, \ldots, p)$ and $(S_1, \ldots, S_p)$ is independent of $(Z_1, \ldots, Z_p)$. In fact, we have

$$\begin{aligned} T_{LH} &= X_1 + \cdots + X_p, \\ T_{LR} &= n \log\left(1 + \frac{1}{n} X_1\right) + \cdots + n \log\left(1 + \frac{1}{n} X_p\right). \end{aligned}$$

Let $Y_i = S_i,\ i = 1, \ldots, p$. Then $Y_1, \ldots, Y_p$ are defined as follows:

(1) For T_{LH}: $Y_1 > \cdots > Y_p > 0$ are the characteristic roots of $\mathbf{W}$ such that $n\mathbf{W} \sim \mathrm{W}_p(n, \mathbf{I}_p)$,
(2) For T_{LR}: $Y_i \sim \chi_{m_i}^2$, $m_i = n - (i - 1)$, $(i = 1, \ldots, p)$, and they are independent.

The above result on T_{LH} is obtained as follows. It is well known that T_{LH} can be expressed as

$$\begin{aligned} T_{LH} &= n\mathrm{tr}(\mathbf{U}'\mathbf{U})\mathbf{S}_e^{-1} \\ &= n\mathrm{tr}(\mathbf{H}'\mathbf{U}'\mathbf{U}\mathbf{H})(\mathbf{H}'\mathbf{S}_e\mathbf{H})^{-1}, \end{aligned}$$

where $\mathbf{U}$ is a $q \times p$ random matrix whose elements are independent identically distributed as N(0, 1), and $\mathbf{H}$ is an orthogonal matrix. Note that the distributions of $\mathbf{H}'\mathbf{U}\mathbf{H}$ and $\mathbf{H}'\mathbf{S}_e\mathbf{H}$ are the same as $\mathbf{U}$ and $\mathbf{S}_e$, respectively. The result is obtained by choosing $\mathbf{H}$ such that $\mathbf{H}'\mathbf{S}_e\mathbf{H} = \mathrm{diag}(Y_1, \ldots, Y_p)$.

For a scale mixture expression of T_{LR}, note that $\Lambda = |\mathbf{S}_e|/|\mathbf{S}_e + \mathbf{S}_h|$ is distributed as a Lambda distribution $\Lambda_p(q, n)$, which, in turn, as it is well known, has identical distribution as a product of p independent beta variables:

$$\Lambda_p(q, n) = \prod_{i=1}^{p} V_i \quad \text{with} \quad V_i \sim \beta\left(\frac{n - i + 1}{2}, \frac{q}{2}\right). \tag{3.9}$$

Therefore, we can write

$$T_{LR} = -n \sum_{i=1}^{p} \log V_i,$$

and V_i can be put in the form

$$V_i = \frac{\chi^2_{m_i}}{\chi^2_{m_i} + \chi^2_q} = \left\{1 + \frac{1}{n} \cdot (\chi^2_{m_i}/n)^{-1} Z_i\right\}^{-1},$$

which implies the assertion with $S_i = \chi^2_{m_i}/n, i = 1, \ldots, p$, where $Z_1, \ldots, Z_p \sim$ $i.i.d.\chi^2_q$.

In the following we assume (3.7). Then, the density function of $\boldsymbol{X}$ can be written as

$$f(\boldsymbol{x}) = \mathrm{E}_{\boldsymbol{Y}}[Y_1 g(xY_1) \cdots Y_p g(xY_p)],$$

where $\boldsymbol{Y} = (Y_1, \ldots, Y_p)'$ and g is the density function of χ^2_q-variate. We consider an asymptotic expansion of $f(\boldsymbol{x})$ given by

$$\begin{aligned} g_{k,p}(\boldsymbol{x}) &= \mathrm{E}\left[g_p(\boldsymbol{x}) + \sum_{j=1}^{k-1} \frac{1}{j!} \left((Y_1 - 1)\frac{\partial}{\partial y_1} + \cdots + (Y_p - 1)\frac{\partial}{\partial y_p} \right)^j \right. \\ &\quad \left. \times\, y_1 g(x_1 y_1) \cdots y_p g(x_p y_p) \Big|_{y_1 = \cdots = y_p = 1} \right] \\ &= g_p(\boldsymbol{x}) + \sum_{j=1}^{k-1} \sum_{(j)} \frac{1}{j_1! \cdots j_p!} b_{j_1}(x_1) \ldots b_{j_p}(x_p) g_p(\boldsymbol{x}) \\ &\quad \times \mathrm{E}\left[(Y_1 - 1)^{j_1} \cdots (Y_p - 1)^{j_p}\right], \end{aligned} \tag{3.10}$$

where $g_p(\boldsymbol{x}) = \prod_{j=1}^{p} g(x_j)$, and the sum $\sum_{(j)}$ is taken over all nonnegative integers such that $j_1 + \cdots + j_p = j$. The function $b_j(x)$ is defined by

$$\frac{\partial^j}{\partial y^j} \{y g(xy)\}\Big|_{y=1} = b_j(x) g(x).$$

For $k = 1, 2;\ b_1(x) = -\frac{1}{2}(x - q),\quad b_2(x) = \frac{1}{4}\{x^2 - 2qx + q(q-2)\}$.

One of the crucial results is to use an error bound for approximation based on $g_{k,p}(\boldsymbol{x})$ in (3.10) which is given as follows. Let $\boldsymbol{X}$ be a general mixtured variate given by (3.6). Then, the density function can be written as

$$f(\boldsymbol{x}) = \mathrm{E}_{\boldsymbol{Y}}\left[Y_1^{-\delta/\rho} g(xY_1^{-\delta/\rho}) \cdots Y_p^{-\delta/\rho} g(xY_p^{-\delta/\rho})\right],$$

where $\boldsymbol{Y} = (Y_1, \ldots, Y_p)'$ and g is the density function of X_i. We have an asymptotic expansion of $f(\boldsymbol{x})$ given by

$$
\begin{aligned}
g_{\delta,k,p}(\boldsymbol{x}) = g_p(\boldsymbol{x}) + \sum_{j=1}^{k-1}\sum_{(j)} \frac{1}{j_1!\cdots j_p!} b_{\delta,j_1}(x_1)\cdots b_{\delta,j_p}(x_p) g_p(\boldsymbol{x}) \\
\times \mathrm{E}\left[(Y_1-1)^{j_1}\cdots(Y_p-1)^{j_p}\right],
\end{aligned} \tag{3.11}
$$

where $g_p(\boldsymbol{x}) = \prod_{j=1}^{p} g(x_j)$, and the sum $\sum_{(j)}$ is taken over all nonnegative integers such that $j_1 + \cdots + j_p = j$. The function $b_j(x)$ is defined by

$$
\frac{\partial^j}{\partial y^j}\left\{y^{-\delta/\rho} g(xy^{-\delta/\rho})\right\}\Bigg|_{y=1} = b_{\delta,j}(x)g(x).
$$

For the approximation (3.11), the following L_1-error bound is given in [2].

Theorem 3.1 *Let $\boldsymbol{X} = (X_1, \ldots, X_p)'$ be a multivariate scale mixture, where $X_i = S_i^{\delta} Z_i, i = 1, \ldots, p,\ \ Z_1, \ldots, Z_p \sim i.i.d.$ with the density $g(x)$ and $Y_i = S_i^{\rho},\ i = 1, \ldots, p$. Suppose that $\mathrm{E}(Y_i^k) < \infty,\ i = 1, \ldots, p$. Then, for any Borel set $A \subset \mathbb{R}^p$, it holds that*

$$
\left|\Pr(\boldsymbol{X} \in A) - \int_A g_{\delta,k,p}(\boldsymbol{x})d\boldsymbol{x}\right| \le \frac{1}{2}\nu_{\delta,k,p}\sum_{i=1}^{p}\mathrm{E}\left[|Y_i - 1|^k\right], \tag{3.12}
$$

where $\nu_{\delta,k,p}$ are determined recursively by the relation $k \ge 2$

$$
\nu_{\delta,k,p} = p^{-1}\left\{\eta_{\delta,k} + (p-1)\sum_{q=0}^{k-1}\nu_{\delta,k-q,p-1}\xi_{\delta,q}\right\}, \quad \text{for } k \ge 2, \tag{3.13}
$$

with $\nu_{\delta,1,p} = \eta_1, \nu_{\delta,k,0} = 0$, and $\nu_{\delta,k,1} = \eta_{\delta,k}$ for all $k \ge 1$, *and*

$$
\xi_{\delta,j} = \frac{1}{j!}\int_0^{\infty}|b_{\delta,j}(x)g(x)|dx, \quad \eta_{\delta,k} = \left\{\xi_{\delta,k}^{1/k} + \left(2 + \sum_{j=1}^{k-1}\xi_{\delta,j}\right)^{1/k}\right\}^k. \tag{3.14}
$$

In the special case $X_i = S_i^{-1}Z_i, i = 1, \ldots, p,\ \ Z_1, \ldots, Z_p \sim i.i.d.$ and $Y_i = S_i,\ i = 1, \ldots, p$, the error bound is expressed as follows: For any Borel set $A \subset \mathbb{R}^p$, it holds that

$$
\left|\mathrm{P}(\boldsymbol{X} \in A) - \int_A g_{k,p}(\boldsymbol{x})d\boldsymbol{x}\right| \le \frac{1}{2}\nu_{k,p}\sum_{i=1}^{p}\mathrm{E}\left[|Y_i - 1|^k\right], \tag{3.15}
$$

where $\nu_{k,p}$ are determined recursively by the relation $k \ge 2$

$$\nu_{k,p} = p^{-1}\left\{\eta_k + (p-1)\sum_{q=0}^{k-1}\nu_{k-q,p-1}\xi_q\right\}, \quad \text{for } k \geq 2, \tag{3.16}$$

with $\nu_{1,p} = \eta_1$, $\nu_{k,0} = 0$, and $\nu_{k,1} = \eta_k$ for all $k \geq 1$, and

$$\xi_j = \frac{1}{j!}\int_0^\infty |b_j(x)g(x)|dx, \quad \eta_k = \left\{\xi_k^{1/k} + \left(2 + \sum_{j=1}^{k-1}\xi_j\right)^{1/k}\right\}^k. \tag{3.17}$$

The distribution functions T_{LH} and T_{LR} are expressed as

$$\Pr(T \leq x) = \int_A f(\boldsymbol{x})d\boldsymbol{x},$$

where the set A is defined as follows:

For $T = T_{LH} : A_x = \{(x_1, \ldots, x_p) \in \mathbb{R}^p : \sum_{i=1}^p x_i \leq x\}$,
For $T = T_{LR} : \tilde{A}_x = \{(x_1, \ldots, x_p) \in \mathbb{R}^p : n\sum_{i=1}^p \log(1 + x_i/n) \leq x\}$.

3.3 Error Bounds for Approximations of T_{LH} and T_{LR}

First we obtain the bounds $B_{1,p}(q, n)$ for T_{LH} as in (3.5) by simplifying (3.15) with $k = 2$. Let

$$M_{r,p} = \mathrm{E}\left[\sum_{i=1}^p (Y_i - 1)^r\right], \quad r = 1, 2, \ldots. \tag{3.18}$$

In this case, Y_i's are the characteristic roots of $\mathbf{W}$; $n\mathbf{W} \sim \mathrm{W}_p(n, \mathbf{I}_p)$,

$$M_{1,p} = \mathrm{E}\left[\mathrm{tr}(\mathbf{W} - \mathbf{I}_p)\right] = 0, \quad M_{2,p} = \mathrm{E}\left[\mathrm{tr}(\mathbf{W} - \mathbf{I}_p)^2\right] = \frac{1}{n}p(p+1),$$

and we have the following theorem.

Theorem 3.2 *Under the assumption $n \geq p$, it holds that*

$$|\Pr(T_{LH} \leq x) - G_r(x)| \leq \frac{1}{2n}p(p+1)\nu_{2,p}, \tag{3.19}$$

where $\nu_{2,p} = \eta_2 + \frac{1}{2}(p-1)\xi_1\eta_1$, and η_k, $k = 1, 2$ are defined by (3.17).

Next we consider the bound $B_{1,p}(q,n)$ for T_{RL}. From (3.15) we have

$$|\Pr(T_{LR} \le x) - G_r(x)| \le B_{11} + B_{12} + B_{13}, \tag{3.20}$$

where

$$B_{11} = \frac{1}{2}\nu_{2,p}\sum_{i=1}^{p} \mathrm{E}[|Y_i - 1|^2], \quad B_{13} = \left|\int_{\tilde{A}_x} g_p(\boldsymbol{x})d\boldsymbol{x} - G_r(x)\right|,$$

$$B_{12} = \left|\int_{\tilde{A}_x} \sum_{i=1}^{p} b_1(x) g_p(\boldsymbol{x}) d\boldsymbol{x} \mathrm{E}[(Y_i - 1)]\right|.$$

In this case, $Y_i = \chi^2_{m_i}$, $i = 1, \ldots, p$, and we have

$$M_{1,p} = -\frac{1}{2n}p(p+1), \quad M_{2,p} = \frac{2}{n}p\left\{1 + \frac{1}{12n}(p-1)(2p-7)\right\}. \tag{3.21}$$

It is easy to see that $B_{12} \le \alpha_1 |M_{1,p}|$, where $\alpha_1 = \sup_x |c_1(x)g(x)|$, $c_1(x) = x$, and $g(x)$ is the density function of χ^2_q-variate. Further, it is shown (see, [3]) that $B_{13} \le (p/n)d_p$, where

$$d_p = \frac{1}{4}q(q+2)\left[\left\{1 - \frac{1}{n}q\right\}^{-(q+2)/2} + \left\{1 - \frac{1}{n}(q+2)\right\}^{-(q+4)/2}\right]. \tag{3.22}$$

These imply the following theorem.

Theorem 3.3 *For* $n > \max(p-1, q+2)$, *it holds that*

$$|\Pr(T_{LR} \le x) - G_r(x)| \le \frac{1}{2}\nu_{2,p} M_{2,p} + \frac{p}{n} d_p(q,n) + \alpha_1 \left|M_{1,p}\right|,$$

where $\nu_{2,p} = \eta_2 + \frac{1}{2}(p-1)\xi_1\eta_1$, *and* η_k, $k = 1, 2$ *are defined by* (3.17), $M_{i,p}$, $i = 1, 2$ *are given by* (3.21), *and* d_p *is defined by* (3.22).

Similar result can be obtained for bounds $B_{2,p}(q,n)$ in the case of T_{LH} and T_{LR}.

3.4 Error Bound for T_{BNP}

In this section, we obtain an error bound for χ^2_r-approximation of T_{BNP} by using a relation between T_{BNP} and T_{LH}, where $r = pq$. In the following we give an outline for derivation due to [4]. First we prepare some preliminary results. From Theorem 3.2 we have

$$|\Pr(T_{LH} < x) - G_r(x)| \le \frac{A_r}{n}, \tag{3.23}$$

where $G_r(x)$ is the distribution function of χ_r^2, and $A_r = (1/2)p(p+1)\nu_{2,p}$.

Lemma 3.1 *Let χ_r^2 be a random variable following the chi-square distribution with r degrees of freedom. Then, for every x with $x > r$,*

$$\Pr(\chi_r^2 > x) < c_r(1 - \Phi(w(x))), \tag{3.24}$$

where

$$w(x) = \left(x - r - r\log\frac{x}{r}\right)^{1/2}, \quad c_r = \frac{(2\pi)^{1/2}(r/2)^{(r-1)/2}\mathrm{e}^{-r/2}}{\Gamma(r/2)},$$

and $\Phi(\cdot)$ denotes the distribution function of the standard normal distribution. Moreover, if $r \ge 1$ and $x \ge 25r$, then

$$\Pr(\chi_r^2 > x) < 0.016c_r\mathrm{e}^{-x/6}. \tag{3.25}$$

The result (3.24) was obtained by [10]. The result (3.25) was obtained by [4]. Using (3.23) and (3.25), we have

$$\Pr(T_{LH} \le n) \ge 1 - 0.016c_r\mathrm{e}^{-n/6} - n^{-1}A_r.$$

Note that $T_{LH} = n\mathrm{tr}\mathbf{S}_h\mathbf{S}_e^{-1}$ and $T_{BNP} = n\mathrm{tr}\mathbf{S}_h(\mathbf{S}_h + \mathbf{S}_e)^{-1}$. It is easy to see that $T_{BNP} \le T_{LH}$. Further, when $T_{LH} \le n$, or $\mathbf{S}_h$ and $\mathbf{S}_e$ are such that $T_{LH} \le n$, we have $T_{LH} - T_{LH}^2/n \le T_{BNP} \le T_{LH}$. Then, we can obtain

$$\begin{aligned}\Pr(T_{LH} \le x) &\le \Pr(T_{BNP} \le x)\\ &\le \Pr(T_{LH} - T_{LH}^2/n < x) + 2\Pr(B_n^c).\end{aligned}$$

Approximating $|\mathrm{P}(T_{LH} - T_{LH}^2/n < x) - G_r(x)|$, we get the following result.

Theorem 3.4 *For $n > 100r$, it holds that*

$$|\Pr(T_{BNP} \le x) - G_r(x)| \le \frac{1}{n}D_r,$$

where

$$D_r = 4A_r + 0.45c_r + \frac{2}{\Gamma(r/2)}\left(\frac{r+2}{\mathrm{e}}\right)^{(r+2)/2},$$

and A_r is defined by (3.23).

3.5 Error Bounds for High-Dimensional Approximations

As approximations in a high-dimensional case, consider asymptotic null distributions of the three test statistics under a high-dimensional framework; $p/n \to c \in (0, 1)$ and fixed q, and their error bounds. This assumption implies that $m = n - p + q$ also tends to infinity. The test statistics can be normalized such that their limiting distributions are N(0, 1) as follows:

$$\widetilde{T}_{LR} = -\frac{\sqrt{p}}{\sigma}\left(1 + \frac{m}{p}\right)\left\{\log\frac{|\mathbf{S}_e|}{|\mathbf{S}_e + \mathbf{S}_h|} + q\log\left(1 + \frac{p}{m}\right)\right\},$$

$$\widetilde{T}_{LH} = \frac{\sqrt{p}}{\sigma}\left(\frac{m}{p}\mathrm{tr}\mathbf{S}_h\mathbf{S}_e^{-1} - q\right),$$

$$\widetilde{T}_{BNP} = \frac{\sqrt{p}}{\sigma}\left(1 + \frac{p}{m}\right)\left\{\left(1 + \frac{m}{p}\right)\mathrm{tr}\mathbf{S}_h(\mathbf{S}_e + \mathbf{S}_h)^{-1} - q\right\},$$

where $\sigma = \sqrt{2q(1+r)}$. To make the statements sweeping, let us use $\widetilde{T}_G$ for $G =$ LR, LH, and BNP.

By method of characteristic functions in [9] the following approximation was derived:

$$\Pr(\widetilde{T}_G < z) = \Phi(z) - \phi(z)\left[\frac{1}{\sqrt{p}}\left\{\frac{1}{\sigma}b_1 + \frac{1}{\sigma^3}b_3 H_2(z)\right\}\right. \tag{3.26}$$
$$\left. + \frac{1}{p}\left\{\frac{1}{\sigma^2}b_2 H_1(z) + \frac{1}{\sigma^4}b_4 H_3(z) + \frac{1}{\sigma^6}b_6 H_5(z)\right\}\right] + \mathrm{O}\left(\frac{1}{p\sqrt{p}}\right),$$

where $\Phi(z)$ and $\phi(z)$ are distribution function and probability density function of the standard normal law, respectively; $H_i(z)$ are the Hermite polynomials and $b_i = b_i(r, q)$ are some polynomials in r and q.

Yet the results in [9] were asymptotic: the upper bounds for the remainder terms were not obtained. To date, non-asymptotic results are known only in the case of approximations of $\widetilde{T}_G$ by the standard normal distribution. Namely, in [8] it was proved that

$$|\Pr(\widetilde{T}_{LR} < z) - \Phi(z)| \le C_1(r, q)/\sqrt{p}, \tag{3.27}$$

where $C_1(r, q)$ depends on r and q but does not depend on p. Moreover, the computable exact expression for $C_1(r, q)$ is known.

Furthermore, in [5, 6] it was proved that

$$|\Pr(\widetilde{T}_G < z) - \Phi(z)| \le C_2(r, q)\frac{\log p}{\sqrt{p}}, \tag{3.28}$$

where $C_2(r, q)$ depends on r and q but does not depend on p. Moreover, the computable exact expression, albeit cumbersome, for $C_2(r, q)$ is known.

There are still open problems to remove $\log p$ in the right-hand side of (3.28) and to get non-asymptotic results for approximation of distribution of $\widetilde{T}_G$ by short asymptotic expansions, that is to get upper bounds for remainder term in (3.26).

The proof of (3.27) is based on representation (3.9) and on the Berry–Esseen type bound for normal approximation of F-distribution, see Theorem 2.1 in [8]:

Theorem 3.5 *Let χ_p^2 and χ_n^2 be independent random variables having chi-square distributions with p and n degrees of freedom. Put*

$$A_{n,p} = \left(\frac{np}{2(n+p)}\right)^{1/2}.$$

Then for all $n \geq 1$ and $p \geq 1$ one has

$$\sup_x \left| \Pr\left(A_{n,p} \left(\frac{\chi_p^2/p}{\chi_n^2/n} - 1 \right) \leq x \right) - \Phi(x) \right| \leq C_3 \left(\frac{1}{n} + \frac{1}{p} \right)^{1/2},$$

with $C_3 \leq 6.1323$.

Moreover, in the proofs of (3.27) and (3.28) the following statement is used which is of independent interest.

Lemma 3.2 *Let the random variables $X_1, \ldots, X_k$ be independent and the inequalities $|\Pr(X_j \leq x) - \Phi(x)| \leq D_j$ hold for all x and $j = 1, \ldots, k$ with some constants $D_1, \ldots, D_k$. Then*

$$\left| \Pr\left(\sum_{j=1}^{k} c_j X_j \leq x \right) - \Phi(x) \right| \leq \sum_{j=1}^{k} D_j,$$

where $c_1, \ldots, c_k$ are any constants such that $c_1^2 + \cdots + c_k^2 = 1$.

For the proof of Lemma 3.2, see, e.g., Theorem 3.1 in [8].

References

1. Chandra, T. K., & Ghosh, G. K. (1980). Valid asymptotic expansions for the likelihood ratio statistic and other perturbed chi-square variables. *Sankhyā Series A, 41*, 22–47.
2. Fujikoshi, Y., Ulyanov, V. V., & Shimizu, R. (2005). L_1-norm error bounds for asymptotic expansions of multivariate scale mixtures and their applications to Hotelling's generalized T_0^2. *Journal of Multivariate Analysis, 96*, 1–19.
3. Fujikoshi, Y., & Ulyanov, V. V. (2006). On accuracy of asymptotic expansions for Wilks' lambda distribution. *Journal of Multivariate Analysis, 97*, 1941–1957.
4. Lipatiev, A. A., & Ulyanov, V. V. (2017). On computable estimates for accuracy of approximation for the Bartlett–Nanda–Pillai statistic. *Siberian Advances in Mathematics, 19*, 153–158.

5. Lipatiev, A. A., & Ulyanov, V. V. (2019). Non-asymptotic analysis of Lawley–Hotelling Statistic for high-dimensional data (Russian). *Zapiski Nauchnykh Seminarov S.-Peterburg Otdel. Mat. Inst. Steklov. POMI*, *486*, 178–189.
6. Lipatiev, A. A., & Ulyanov, V. V. (2020). *Non-asymptotic analysis of approximations for Lawley–Hotelling and Bartlett–Nanda–Pillai statistics in high-dimensional settings*. Hiroshima Statistical Research Group, TR, 20-01.
7. Siotani, M., Hayakawa, T., & Fujikoshi, Y. (1985). *Modern multivariate statistical analysis: A graduate course and handbook*. Columbus: American Sciences Press.
8. Ulyanov, V. V., Wakaki, H., & Fujikoshi, Y. (2006). Berry–Esseen bound for high dimensional asymptotic approximation of Wilks' lambda distribution. *Statistics & Probability Letters*, *76*, 1191–1200.
9. Wakaki, H., Fujikoshi, Y., & Ulyanov, V. V. (2014). Asymptotic expansions of the distributions of MANOVA test statistics when the dimension is large. *Hiroshima Mathematical Journal*, *44*, 247–259.
10. Wallace, D. L. (1959). Bounds on normal approximations to Student's and the chi-square distributions. *The Annals of Mathematical Statistics*, *30*, 1121–1130.

Chapter 4
Linear and Quadratic Discriminant Functions

Abstract This chapter is concerned with theoretical accuracies for asymptotic approximations of the expected probabilities of misclassification (EPMC) when the linear discriminant function and the quadratic discriminant function are used. The method in this chapter is based on asymptotic bounds for asymptotic approximations of a location and scale mixture. The asymptotic approximations considered in detail are those in which both the sample size and the dimension are large, and the sample size is large.

4.1 Introduction

An important interest in discriminant analysis is to classify a $p \times 1$ observation vector $\boldsymbol{X}$ as coming from one of two populations Π_1 and Π_2. Let Π_i be the two p-variate normal population with $\mathrm{N}_p(\boldsymbol{\mu}_i, \boldsymbol{\Sigma})$, where $\boldsymbol{\mu}_1 \neq \boldsymbol{\mu}_2$ and $\boldsymbol{\Sigma}$ is positive definite. When the values of the parameters are unknown, assume that random samples of sizes N_1 and N_2 are available from Π_1 and Π_2, respectively. Let $\bar{\boldsymbol{X}}_1$, $\bar{\boldsymbol{X}}_2$, and $\mathbf{S}$ be the sample mean vectors and the sample covariance matrix. Then, a well-known linear discriminant function is defined by

$$W = (\bar{\boldsymbol{X}}_1 - \bar{\boldsymbol{X}}_2)'\mathbf{S}^{-1}\left\{\boldsymbol{X} - \frac{1}{2}(\bar{\boldsymbol{X}}_1 + \bar{\boldsymbol{X}}_2)\right\}. \tag{4.1}$$

The observation $\boldsymbol{X}$ may be classified to Π_1 or Π_2 according to $W \geq 0$ or $W < 0$. We also have a quadratic discriminant function Q defined by

$$\begin{aligned} Q = \frac{1}{2}&\{(1 + N_2^{-1})^{-1}(\boldsymbol{X} - \bar{\boldsymbol{X}}_2)'\mathbf{S}^{-1}(\boldsymbol{X} - \bar{\boldsymbol{X}}_2) \\ &-(1 + N_1^{-1})^{-1}(\boldsymbol{X} - \bar{\boldsymbol{X}}_1)'\mathbf{S}^{-1}(\boldsymbol{X} - \bar{\boldsymbol{X}}_1)\}. \end{aligned} \tag{4.2}$$

Then the observation $\boldsymbol{X}$ is usually classified to Π_1 or Π_2 according to $Q \geq 0$ or $Q < 0$.

Y. Fujikoshi and V. V. Ulyanov, *Non-Asymptotic Analysis of Approximations for Multivariate Statistics*, JSS Research Series in Statistics,
https://doi.org/10.1007/978-981-13-2616-5_4

There are two types of misclassification probability. One is the probability of allocating $\boldsymbol{x}$ to Π_2 even though it actually belongs to Π_1. The other is the probability that $\boldsymbol{x}$ is classified as Π_1 although it actually belongs to Π_2. It is important to evaluate the expected probabilities of misclassification (EPMC):

$$e_W(2|1) = \Pr(W \leq 0 \mid \boldsymbol{X} \in \Pi_1), \quad e_W(1|2) = \Pr(W \geq 0 \mid \boldsymbol{X} \in \Pi_2), \tag{4.3}$$

$$e_Q(2|1) = \Pr(Q \leq 0 \mid \boldsymbol{X} \in \Pi_1), \quad e_Q(1|2) = \Pr(Q \geq 0 \mid \boldsymbol{X} \in \Pi_2). \tag{4.4}$$

As is well known, the distribution of W or Q when $\boldsymbol{X} \in \Pi_1$ is the same as that of $-W$ (or $-Q$) when $\boldsymbol{X} \in \Pi_2$. This indicates that $e_W(1|2)$ (or $e_Q(1|2)$) is obtained from $e_W(2|1)$ (or $e_Q(2|1)$) by replacing (N_1, N_2) with (N_2, N_1). Thus, in this chapter, we only deal with $e_W(2|1)$ and $e_Q(2|1)$.

In general, exact evaluation of EPMC is cumbersome, while there is a considerable literature concerning their asymptotic approximations including asymptotic expansions. It should be noted that there are two types (type-I, type-II) of asymptotic approximations. Type-I approximations are the ones under a framework such that N_1 and N_2 are large and p is fixed. For a review of these results, see, e.g., [11]. Naturally, the accuracy of type-I approximations will become bad as the dimension p is large. On the other hand, type-II approximations are the ones under a framework such that N_1, N_2, and p are large. It has been noted (e.g., [5, 13]) that type-II approximations by [5, 9] have superior overall accuracy in comparison with some other approximations including a type-I asymptotic expansion by [8]. The purpose of the present chapter is to obtain error bounds of type-II asymptotic approximations for the W-rules and Q-rule. Their error bounds are obtained by establishing a general approximation result based on a location and scale mixture. It is noted that the general approximation result can also be applied to a theoretical accuracy of type-I approximations and some related distributions.

4.2 Location and Scale Mixture Expression for EPMC

In this section, we shall see that the distributions of W and Q are expressed as location and scale mixtures of the standardized normal distribution. In general, a random variable X is called a location and scale mixture of the standardized normal distribution, if X is expressed as

$$X = V^{1/2}Z - U, \tag{4.5}$$

where $Z \sim \mathrm{N}(0, 1)$, Z and (U, V) are independent, and $V > 0$. It is known (see [2]) that the linear discriminant function W can be expressed as a location and scale mixture of the standardized normal distribution as

$$W = V^{1/2}Z - U. \tag{4.6}$$

Here, when $\boldsymbol{X}$ comes from Π_1, the variables (Z, U, V) may be defined as

$$
\begin{aligned}
Z &= V^{-\frac{1}{2}}(\bar{\boldsymbol{X}}_1 - \bar{\boldsymbol{X}}_2)'\mathbf{S}^{-1}(\boldsymbol{X} - \boldsymbol{\mu}_1), \\
U &= (\bar{\boldsymbol{X}}_1 - \bar{\boldsymbol{X}}_2)'\mathbf{S}^{-1}(\bar{\boldsymbol{X}}_1 - \boldsymbol{\mu}_1) - \frac{1}{2}D^2, \\
V &= (\bar{\boldsymbol{X}}_1 - \bar{\boldsymbol{X}}_2)'\mathbf{S}^{-1}\boldsymbol{\Sigma}\mathbf{S}^{-1}(\bar{\boldsymbol{X}}_1 - \bar{\boldsymbol{X}}_2),
\end{aligned}
\tag{4.7}
$$

where $D = \left\{(\bar{\boldsymbol{X}}_1 - \bar{\boldsymbol{X}}_2)'\mathbf{S}^{-1}(\bar{\boldsymbol{X}}_1 - \bar{\boldsymbol{X}}_2)\right\}^{1/2}$ is the sample Mahalanobis distance between two populations. This property of W has been used in [7].

As an alternative, consider the quadratic discriminant function Q in (4.2). Letting $a = (1 + N_1^{-1})^{-1}(1 + N_2^{-1})$, we can state

$$
\begin{aligned}
Q &= \frac{1}{2}(1 + N_2^{-1})^{-1}\left\{-\sqrt{a}(\boldsymbol{X} - \bar{\boldsymbol{X}}_1) + \boldsymbol{X} - \bar{\boldsymbol{X}}_2\right\}'\mathbf{S}^{-1} \\
&\qquad \times \left\{\sqrt{a}(\boldsymbol{X} - \bar{\boldsymbol{X}}_1) + \boldsymbol{X} - \bar{\boldsymbol{X}}_2\right\} && (4.8) \\
&= \frac{1}{2}(1 + N_2^{-1})^{-1}b_1 b_2 \boldsymbol{T}_1'\mathbf{B}^{-1}\boldsymbol{T}_2, && (4.9)
\end{aligned}
$$

where

$$
\begin{aligned}
\boldsymbol{T}_1 &= b_1^{-1}\boldsymbol{\Sigma}^{-1/2}\left\{(-\sqrt{a} + 1)\boldsymbol{x} + \sqrt{a}\bar{\boldsymbol{X}}_1 - \bar{\boldsymbol{X}}_2\right\}, \\
\boldsymbol{T}_2 &= b_2^{-1}\boldsymbol{\Sigma}^{-1/2}\left\{(\sqrt{a} + 1)\boldsymbol{X} - \sqrt{a}\bar{\boldsymbol{X}}_1 - \bar{\boldsymbol{X}}_2\right\}, \\
\mathbf{B} &= \boldsymbol{\Sigma}^{-1/2}\mathbf{S}\boldsymbol{\Sigma}^{-1/2},
\end{aligned}
\tag{4.10}
$$

and

$$
b_1 = \sqrt{2}\left\{1 + N_2^{-1} - \sqrt{a}\right\}^{1/2}, \quad b_2 = \sqrt{2}\left\{1 + N_2^{-1} + \sqrt{a}\right\}^{1/2}. \tag{4.11}
$$

Suppose that $\boldsymbol{X}$ belongs to Π_1, i.e., $\boldsymbol{X} \sim \mathrm{N}(\boldsymbol{\mu}_1, \boldsymbol{\Sigma})$. Then, $\boldsymbol{T}_i \sim \mathrm{N}(b_i^{-1}\boldsymbol{\delta}, \mathbf{I}_p)$, $i = 1, 2$, $n\mathbf{B} \sim \mathrm{W}_p(n, \mathbf{I}_p)$, and $\boldsymbol{T}_1, \boldsymbol{T}_2$ and $\mathbf{B}$ are independent. Therefore, we have a location and scale expression given by

$$
Q = b\left\{V^{1/2}Z - U\right\}, \tag{4.12}
$$

where

$$
\begin{aligned}
Z &= (\boldsymbol{T}_1\mathbf{B}^{-2}\boldsymbol{T}_1)^{-1/2}\boldsymbol{T}_1'\mathbf{B}^{-1}(\boldsymbol{T}_2 - b_2^{-1}\boldsymbol{\delta}), \\
U &= c_1\boldsymbol{\gamma}'\mathbf{B}^{-1}\boldsymbol{T}_1, \quad V = c_2\boldsymbol{T}_1'\mathbf{B}^{-2}\boldsymbol{T}_1.
\end{aligned}
\tag{4.13}
$$

Here,

$$
\begin{aligned}
&c_1 = -\{N/(N_1N_2)\}^{1/2} b_1 b_2^{-1}, \quad c_2 = \{N/(N_1N_2)\}^{1/2}, \\
&b = \left[(1+N^{-1})/\{(1+N_1^{-1})(1+N_2^{-1})\}\right]^{1/2}, \\
&\gamma = b_1^{-1}\delta, \quad \tau^2 = \gamma'\gamma = b_1^{-2}\Delta^2.
\end{aligned} \tag{4.14}
$$

The location and scale mixture (4.12) was obtained by [4, 12]. A more general quadratic discriminant function is defined by (4.8) with a general positive constant g for a. It is known [4] that the quadratic function can be expressed as a location and scale mixture.

4.3 General Approximation and Error Bounds

In this section, we give a general approximation formula for a location and scale mixture of $X = V^{1/2}Z - U$ of the standard normal distribution, and its error bound. The distribution function can be expressed as

$$
\begin{aligned}
F(x) &= \Pr(X \le x) \\
&= \mathrm{E}_{(U,V)}\left\{\Phi(V^{-1/2}(U+x))\right\}.
\end{aligned} \tag{4.15}
$$

Choosing a pivotal point (u_0, v_0) of (U, V), we can propose an approximation

$$
F(x) \approx \Phi\left(v_0^{-1/2}(u_0+x)\right).
$$

The pivotal point may be any point in the range of (U, V). As a reasonable point, we consider

$$
(u_0, v_0) = (\mathrm{E}(U), \mathrm{E}(V)), \quad (u_0, v_0) = (\lim U, \lim V).
$$

The first choice $(\mathrm{E}(U), \mathrm{E}(V))$ is related to the type-II approximations. The second choice $(\lim U, \lim V)$ under a large sample is related to the type-I approximations.

Now, let us consider a more refinement approximation of

$$
G(u, v) = \Phi\left(v^{-1/2}(u+x)\right) \tag{4.16}
$$

by considering its expansion around (u_0, v_0). First we expand $G(u, v)$ with respect to v at $v = v_0$ as

$$
G(u, v) = \sum_{j=0}^{k-1} \frac{1}{j!}\left\{\frac{\partial^j}{\partial v^j} G(u, v)\right\}_{v=v_0} (v-v_0)^j + R_{k,k}, \tag{4.17}
$$

where

$$R_{k,k} = \frac{1}{k!}\left\{\frac{\partial^k}{\partial v^k}G(u,v)\right\}_{v=v^*}(v-v_0)^k \tag{4.18}$$

and $v^* \in (v_0, v)$ or (v, v_0). Next, consider Taylor's expansions for $j = 0, 1, \ldots, k-1$;

$$\begin{aligned}\left\{\frac{\partial^j}{\partial v^j}G(u,v)\right\}_{v=v_0} &= \sum_{i=0}^{k-j-1}\frac{1}{i!}\left[\frac{\partial}{\partial u^i}\left\{\frac{\partial^j}{\partial v^j}G(u,v)\right\}_{v=v_0}\right]_{u=u_0}(u-u_0)^i \\ &+\frac{1}{(k-j)!}\left[\frac{\partial^{k-j}}{\partial u^{k-j}}\left\{\frac{\partial^j}{\partial v^j}G(u,v)\right\}_{v=v_0}\right]_{u=u_j^*}(u-u_0)^{k-j},\end{aligned}$$

where $u_j^* \in (u, u_0)$ or (u_0, u). Substituting the above expressions into (4.17) and rearranging the resulting expressions, we obtain

$$G(u,v) = G_k(u,v) + \sum_{j=0}^{k} R_{k,j}, \tag{4.19}$$

where

$$\begin{aligned}G_k(u,v) &= \sum_{j=0}^{k-1}\sum_{i=0}^{j}\frac{1}{i!(j-i)!}c_{i,j-i}(u_0,v_0)(u-u_0)^i(v-v_0)^{j-i}, \\ R_{k,j} &= \frac{1}{j!(k-j)!}\left[\frac{\partial^j}{\partial u^j}\left\{\frac{\partial^{k-j}}{\partial v^{k-j}}G(u,v)\right\}_{v=v_0}\right]_{u=u_j^*} \\ &\quad\times(u-u_0)^j(v-v_0)^{k-j}, \quad (j=0,1,\ldots,k-1), \\ c_{i,j-i}(u_0,v_0) &= \left[\frac{\partial^i}{\partial u^i}\left\{\frac{\partial^{j-i}}{\partial v^{j-i}}G(u,v)\right\}_{v=v_0}\right]_{u=u_0}.\end{aligned}$$

Note that $G_k(u, v)$ is not the usual Taylor's expansion up to the $(k-1)$th order.

Now we restrict our attention to $G(u.v) = \Phi\left(v^{-\frac{1}{2}}(u+x)\right)$ on a set $(-\infty, \infty) \times (0, \infty)$, where x is any fixed number. Let H_j be the jth Hermite polynomial (see, Sect. 2.3.1) defined by

$$\frac{\partial^j}{\partial y^j}\phi(y) = (-1)^j H_j(y)\phi(y), \tag{4.20}$$

where ϕ is the density function of N(0, 1). The following Lemmas 4.1 and 4.2 are fundamental in deriving an asymptotic formula and its error bound.

Lemma 4.1

$$\text{(i)}\ \frac{\partial^j}{\partial v^j}\Phi\left(v^{-1/2}(u+x)\right) = -\frac{1}{2^j}H_{2j-1}(y)\phi(y)v^{-j},$$

$$\text{(ii)}\ \left[\frac{\partial^i}{\partial u^i}\left\{\frac{\partial^{j-i}}{\partial v^{j-i}}\Phi\left(v^{-1/2}(u+x)\right)\right\}_{v=v_0}\right]_{u=u_0}$$
$$= -\frac{(-1)^i}{2^{j-i}}H_{2j-i-1}(y_0)\phi(y_0){v_0}^{-(j-i/2)}$$
$$= c_{i,j-i}(u_0, v_0; x), \quad j \geq i,$$

where $c_{0,0}(u_0, v_0; x) = \Phi\left({v_0}^{-1/2}(u_0+x)\right)$, $y = v^{-1/2}(u+x)$ *and* $y_0 = {v_0}^{-1/2} \times (u_0+x)$.

Proof (i) can be proved by induction, using a recurrence relation

$$H_{2j+1}(x) = (x^2-2j)H_{2j-1}(x) - xH'_{2j-1}(x)$$

which follows from two well-known formulas $H_{n+1}(x) = xH_n(x) - H'_n(x)$ and $H_{n+1}(x) - xH_n(x) - nH_{n-1}(x) = 0$. (ii) follows from (i) and (4.19). □

We use the following notation:

$$\begin{aligned} h_j &= \sup|H_j(x)\phi(x)|, \\ \alpha_j &= \begin{cases} \dfrac{1}{2}, & (j=0), \\ \dfrac{1}{2^j j!}h_{2j-1}, & (j=1,2,\ldots k), \end{cases} \\ d_k &= \alpha_0+\alpha_1+\cdots+\alpha_{k-1}, \\ \beta_{k,k-j} &= \begin{cases} ({\alpha_k}^{1/k}+{d_k}^{1/k})^k, & (j=0), \\ \dfrac{1}{j!(k-j)!2^{k-j}}h_{2k-j-1}, & (j=1,\ldots,k). \end{cases} \end{aligned} \tag{4.21}$$

Lemma 4.2 *Let* $R_{k,k-j}$, $j=0,1,\ldots,k$ *be the quantities defined by* (4.17) *and* (4.19) *for* $G(u,v) = \Phi\left(v^{-1/2}(u+x)\right)$. *Then, for* $j = 0, 1, \ldots, k$,

$$|R_{k,k-j}| \leq \beta_{k,k-j}{v_0}^{-(k-j/2)}|u-u_0|^j|v-v_0|^{k-j}. \tag{4.22}$$

Proof For $j = 1, \ldots, k$, from Lemma 4.1(ii) we have

$$
\begin{aligned}
R_{k,k-j} &= -\frac{(-1)^j}{j!(k-j)!2^{k-j}} H_{2k-j-1}(y_j{}^*)\phi(y_j{}^*) \\
&\quad \times v_0{}^{-(k-j/2)}(u-u_0)^j(v-v_0)^{k-j},
\end{aligned}
$$

where $y_j{}^* = v_0{}^{-1/2}(u_j{}^* + x)$. This implies (4.22) for $j = 1, \ldots, k$. So, it is sufficient to show

$$
|R_{k,k}| \le \beta_{k,k} v_0{}^{-k} |v - v_0|^k.
$$

The result in the case where $v_0 = 1$ was proved in [10]. For any $v_0 > 0$ we can modify the proof in the case of $v_0 = 1$ as follows. Let c be a given constant such that $0 < c < v_0$. From (4.18) and Lemma 4.1(i) we have

$$
|R_{k,k}| \le c^{-k}\alpha_k |v - v_0|^k, \quad \text{for} \quad v \ge c.
$$

If $0 < v < c$, from (4.18) and Lemma 4.1(i) we have

$$
\begin{aligned}
|R_{k,k}| &\le \left|\Phi\left(v^{-1/2}\tilde{u}\right) - \Phi\left(v_0{}^{-1/2}\tilde{u}\right)\right| \\
&\quad + \sum_{j=1}^{k-1} \frac{1}{j!2^j} \left|H_{2j-1}\left(v_0{}^{-1/2}\tilde{u}\right)\phi\left(v_0{}^{-1/2}\tilde{u}\right)\right| \left|\frac{v - v_0}{v_0}\right|^j \\
&\le d_k |v - v_0|^{-k} |v - v_0|^k \\
&\le (v_0 - c)^{-k} d_k |v - v_0|^k,
\end{aligned}
$$

where $\tilde{u} = u + x$. The best choice of c is given as a solution of $c^{-k}\alpha_k = (v_0 - c)^{-k} d_k$, i.e., $c = c_* = v_0\left\{1 + (d_k/\alpha_k)^{1/k}\right\}^{-1}$, and then we have $c_*{}^{-k}\alpha_k = (v_0 - c_*)^{-k} d_k = \beta_{k,k} v_0{}^{-k}$. This completes the proof. □

Lemma 4.3 *Let* (u_0, v_0) *be any given point in a set* $(-\infty, \infty) \times (0, \infty)$, *and*

$$
\Phi_k(u, v; x) = \sum_{j=0}^{k-1}\sum_{i=0}^{j} \frac{1}{i!(j-i)!} c_{i,j-i}(u_0, v_0; x)(u - u_0)^i (v - v_0)^{j-i}, \tag{4.23}
$$

where the $c_{i,j-i}$*'s are given by* Lemma 4.1(ii). *Then it holds that*

$$
\begin{aligned}
\left|\Phi\left(v^{-\frac{1}{2}}(u + x)\right) - \Phi_k(u, v; x)\right| &\le \sum_{j=0}^{k} \beta_{k,k-j} v_0{}^{-(k-j/2)} \\
&\quad \times |u - u_0|^j |v - v_0|^{k-j},
\end{aligned} \tag{4.24}
$$

where the $\beta_{k,k-j}$'s are given by (4.21).

Proof The result follows from (4.19) and Lemma 4.2. □

Theorem 4.1 *Let $X = V^{1/2}Z - U$ be a random variable such that $Z \sim \mathrm{N}(0, 1)$, $V > 0$ and Z is independent of (U, V). Suppose that the kth moments of (U, V) exist. For any real numbers $u_0, v_0 > 0$ and x, let*

$$\begin{aligned}\Phi_k(u_0, v_0; x) &= \mathrm{E}\left[\Phi_k(U, V; x)\right] \\ &= \sum_{j=0}^{k-1}\sum_{i=0}^{j}\frac{1}{i!(j-i)!}c_{i,j-i}(u_0, v_0; x) \\ &\quad\times \mathrm{E}\left[(U-u_0)^j (V-v_0)^{j-i}\right],\end{aligned} \tag{4.25}$$

where the $c_{i,j-i}(u_0, v_0; x)$'s are given by Lemma 4.1(ii). *Then*

$$\begin{aligned}|\mathrm{P}(X \le x) - \Phi_k(u_0, v_0; x)| \le \sum_{j=0}^{k}\beta_{k,k-j}{v_0}^{-(k-j/2)} \\ \times \mathrm{E}\left\{|U-u_0|^j|V-v_0|^{k-j}\right\},\end{aligned} \tag{4.26}$$

where the $\beta_{k,k-j}$'s are given by (4.21).

Proof From the definition of X we have

$$\mathrm{P}(X \le x) = \mathrm{E}_{(U,V)}\left\{\Phi\left(V^{-1/2}(U+x)\right)\right\}.$$

Further, from Lemma 4.3 we have

$$\begin{aligned}-\sum_{j=0}^{k}\beta_{k,k-j}v_0^{-(k+j)/2}|U-u_0|^j|V-v_0|^{k-j} \\ \le \Phi\left(V^{-1/2}(U+x)\right) - \Phi_k(U, V; x) \\ \le \sum_{j=0}^{k}\beta_{k,k-j}{v_0}^{-(k-j/2)}|U-u_0|^j|V-v_0|^{k-j}.\end{aligned}$$

The result (4.26) is obtained by considering the expectations of the above inequalities with respect to (U, V). □

Corollary 4.1 *In* Theorem 4.1, *assume that* $\mathrm{E}(U^2) < \infty$ *and* $\mathrm{E}(V^2) < \infty$. *Then*

$$\left|\Pr\{Y \le y\} - \Phi(\tilde{y})\right| \le B_0 + B_1, \tag{4.27}$$

where $\tilde{y} = v_0^{-1/2}(y + u_0)$, *and*

$$\begin{aligned} B_0 &= \beta_{2,0} v_0^{-1} \mathrm{E}[(U - u_0)^2] + \beta_{2,2} v_0^{-2} \mathrm{E}[(V - v_0)^2] \\ &\quad + \beta_{2,1} v_0^{-3/2} \{\mathrm{E}[(U - u_0)^2]\mathrm{E}[(V - v_0)^2]\}^{1/2}, \\ B_1 &= \frac{1}{\sqrt{2\pi}} v_0^{-1/2} |\mathrm{E}(U - u_0)| + \frac{1}{2\sqrt{2\pi e}} v_0^{-1} |\mathrm{E}(V - v_0)|, \end{aligned}$$

where the constants $\beta_{2,j}$ *are expressed in terms of* h_j *as*

$$\beta_{2,0} = \frac{1}{2} h_1, \quad \beta_{2,1} = \frac{1}{2} h_2, \quad \beta_{2,2} = \frac{1}{2} \left\{ \sqrt{1 + h_1} + \frac{1}{2}\sqrt{h_3} \right\}^2. \tag{4.28}$$

Corollary 4.2 *In* Corollary 4.1, *assume that* $u_0 = \mathrm{E}(U)$, *and* $v_0 = \mathrm{E}(V)$. *Then*

$$\left| \Pr\{Y \le y\} - \Phi(\tilde{y}) \right| \le B_0, \tag{4.29}$$

where $\tilde{y} = v_0^{-1/2}(y + u_0)$, *and*

$$\begin{aligned} B_0 &= \beta_{2,0} v_0^{-1} \mathrm{Var}(U) + \beta_{2,0} v_0^{-2} \mathrm{Var}(V) \\ &\quad + \beta_{2,1} v_0^{-3/2} \{\mathrm{Var}(U)\mathrm{Var}(V)\}^{1/2}. \end{aligned}$$

Remark 4.1 From [10] we can replace the constants $\beta_{k,k}$ for $k \le 6$ by $1/2$, and so, $\beta_{2,2}$ can be replaced by an improved constant $1/2$.

4.4 Error Bounds for EPMC

Using Corollaries 4.1 and 4.2, it is possible to obtain asymptotic approximations and their error bounds of expected probabilities of misclassification (EPMC) of W and Q, i.e., $e_W(2|1)$ and $e_Q(2|1)$. The results on $e_W(2|1)$ are given in [4, 6].

In the following we consider $e_Q(2|1)$ which is expressed as

$$e_Q(2|1) = \mathrm{E}_{(U,V)}[\Phi(V^{-1/2}U)], \tag{4.30}$$

where U and V are defined by (4.13). Let us choose the range point (u_0, v_0) of (U, V) as

$$u_0 = \mathrm{E}(U), \quad v_0 = \mathrm{E}(V). \tag{4.31}$$

Consider approximating $e_Q(2|1)$ by $\Phi(v_0^{-1/2}u_0)$. For use of Corollary 4.2, it requires the means and variances of U and V in (4.32) which are given in the following lemma.

Lemma 4.4 *Let U and V be random variables defined by* (4.13). *Then their means and variances are given as follows:*

$$\begin{aligned}
\mathrm{E}(U) &= \frac{nc_1\tau^2}{m-1}, \quad m > 1, \\
\mathrm{Var}(U) &= \frac{(nc_1)^2\tau^2}{(m-1)(m-3)}\left(\frac{n-1}{m} + \frac{2\tau^2}{m-1}\right), \quad m > 3, \\
\mathrm{E}(V) &= \frac{n^2(n-1)(p+\tau^2)}{m(m-1)(m-3)}, \quad m > 3, \\
\mathrm{Var}(V) &= \frac{n^4(n-1)}{m(m-1)(m-3)}\left[\frac{2(n-3)(p+2\tau^2)}{(m-2)(m-5)(m-7)}\right. \\
&\quad \left. + (p+\tau^2)^2\left\{\frac{n-3}{(m-2)(m-5)(m-7)} - \frac{n-1}{m(m-1)(m-3)}\right]\right., \\
&\quad m > 7,
\end{aligned}$$

where c_1 is given by (4.14)*, $m = n - p$, and $\tau^2 = b_1^{-2}\Delta^2$.*

Proof The random variables U and V are expressed as

$$U = nc_1\boldsymbol{\gamma}'\mathbf{A}^{-1}\boldsymbol{T}_1, \quad V = n^2c_2\boldsymbol{T}_1'\mathbf{A}^{-2}\boldsymbol{T}_1, \tag{4.32}$$

where $\mathbf{A} = n\mathbf{B}$, Note that $\boldsymbol{T}_1 \sim \mathrm{N}_p(\boldsymbol{\gamma}, \mathbf{I}_p)$, $\mathbf{A} \sim \mathrm{W}_p(n, \mathbf{I}_p)$, and $\boldsymbol{T}_1$ and $\mathbf{A}$ are independent. The results are obtained by using the following distributional expressions (see, e.g., [3, 12]):

$$\begin{aligned}
\boldsymbol{\gamma}'\mathbf{A}^{-1}\boldsymbol{T}_1 &= \tau Y_1^{-1}\left\{Z_1 + \tau - (Y_2/Y_3)^{1/2}Z_2\right\}, \\
\boldsymbol{T}_1'\mathbf{A}^{-2}\boldsymbol{T}_1 &= Y_1^{-2}\left(1 + Y_2Y_3^{-1}\right)\left\{(Z_1+\tau)^2 + Z_2^2 + Y_4\right\}.
\end{aligned}$$

Here, $Y_i \sim \chi^2_{f_i}$, $i = 1, \ldots 4$; $Z_i \sim \mathrm{N}(0, 1)$, $i = 1, 2$; and

$$f_1 = m + 1, \quad f_2 = p - 1, \quad f_3 = m + 2, \quad f_4 = p - 2.$$

Further, all the variables Y_1, Y_2, Y_3, Y_4, Z_1, and Z_2 are independent. □

Let us consider an approximation

$$e_Q(2|1) \sim \Phi(x_0), \quad x_0 = v_0^{-1/2}u_0, \tag{4.33}$$

where $u_0 = \mathrm{E}(U)$ and $v_0 = \mathrm{E}(V)$. Applying Corollary 4.2 to this approximation, we have the following result. If $m = N_1 + N_2 - p - 2 > 7$,

$$|e_Q(2|1) - \Phi(y_0)| \leq B_0, \tag{4.34}$$

where

$$B_0 = \frac{1}{2\sqrt{2\pi e}} v_0^{-1} V_U + \frac{1}{2} v_0^{-2} V_V + \frac{1}{2\sqrt{2\pi}} v_0^{-3/2} \{V_U V_V\}^{1/2}, \tag{4.35}$$

and $V_U = \mathrm{Var}(U)$ and $V_V = \mathrm{Var}(V)$ are given by Lemma 4.4. Let us consider the high-dimension and large-sample asymptotic framework given by

$$\text{(AF)}: \quad p/N_i \to h_i > 0, \quad i = 1, 2, \quad \Delta^2 = \mathrm{O}(1). \tag{4.36}$$

Then, under (AF), from Corollary 4.2 we have

$$B_0 = \mathrm{O}_1, \quad \text{and} \quad e_Q(2|1) = \Phi(y_0) + \mathrm{O}_1, \tag{4.37}$$

where O_j denotes the term of jth order with respect to $(N_1^{-1}, N_2^{-1}, p^{-1})$. The result (4.34) gives a computable error bound as well as a non-asymptotic formula.

4.5 Some Related Topics

In this section we take up some related topics, focusing on the linear discriminant function W under a large-sample framework; $N_1 \to \infty$, $N_2 \to \infty$, and p is fixed. Note that W is expressed as a location and scale mixture given (4.6) and (4.7). Then, it is easily seen that $U \to -(1/2)\Delta^2$ and $V \to \Delta^2$. Therefore, applying Corollary 4.1 with $(u_0, v_0) = (-(1/2)\Delta^2, \Delta^2)$ to W,

$$\left| \Pr\left\{ \frac{W - \frac{1}{2}\Delta^2}{\Delta} \le x \mid \Pi_1 \right\} - \Phi(x) \right| \le B_W^{(1)}. \tag{4.38}$$

For an explicit expression for $B_W^{(1)}$, see [6].

The distributions of the Studentized W are defined by [1] as the ones of $\{W - (1/2)D^2\}/D$ and $\{W + (1/2)D^2\}/D$ under Π_1 and Π_2, respectively. As a special case, the limiting distributions are the standard normal distribution. So we are interested in obtaining an error bound for

$$\left| \Pr\left\{ \frac{W - \frac{1}{2}D^2}{D} \le x \mid \Pi_1 \right\} - \Phi(x) \right|.$$

Using (4.6) we can write the Studentized statistic W_S of W under Π_1 as a location and scale mixture of the standard normal distribution as follows.

$$W_S \equiv D^{-1}\left(W - \frac{1}{2}D^2 \right) = V^{1/2} Z - U, \tag{4.39}$$

where $U = (\bar{X}^{(1)} - \bar{X}^{(2)})'\mathbf{S}^{-1}(\bar{X}^{(1)} - \mu^{(1)})/D, V = \tilde{D}/D$ and $\widetilde{D}^2 = (\bar{X}^{(1)} - \bar{X}^{(2)})'$ $\mathbf{S}^{-1}\Sigma\mathbf{S}^{-1}(\bar{X}^{(1)} - \bar{X}^{(2)})$. Therefore we can apply Corollary 4.1 with an appropriate (u_0, v_0). In this case we take $(u_0, v_0) = (0, 1)$. Finally we obtain

$$|\Pr\{W_S \leq x \mid X \in \Pi_1\} - \Phi(x)| \leq \widetilde{B}_W^{(1)}. \tag{4.40}$$

For an explicit expression of $\widetilde{B}_W^{(1)}$, see [6]. The bound in (4.40) depends on the sample sizes (N_1, N_2) only, but its order is $O_{1/2}$. It is interesting to find a sharp bound for $|E(U)|$ whose order is O_1. We note that for the Studentized classification statistic, the inequality (4.40) is valid under the conditions $n - p - 3 > 0$. On the other hand, the corresponding condition for the classification statistic W is $n - p - 7 > 0$.

Next, let us consider the quantity $Q_D = \Phi(-D/2)$ which plays an important role in estimating the probabilities of misclassification. The expectation of Q_D converges to $\Phi(-\Delta/2)$. We are interested in the approximation error $|E(Q_D) - \Phi(-\Delta/2)|$ which can be derived by using $E(Q_D) = \Pr\{X \leq -(1/2)\Delta\}$. Here X is a scale mixture of the standard normal distribution defined by $X = V^{1/2}Z$, where Z is the standard normal variate, $V = \Delta^2/D^2$, and Z and V are independent.

References

1. Anderson, T. W. (1973). An asymptotic expansion of the distribution of the studentized classification statistic W. *The Annals of Statistics, 1*, 964–972.
2. Fujikoshi, Y. (2000). Error bounds for asymptotic approximations of the linear discriminant function when the sample size and dimensionality are large. *Journal of Multivariate Analysis, 73*, 1–17.
3. Fujikoshi, Y. (2002). Selection of variables for discriminant analysis in a high-dimensional case. *Sankhyā Series A, 64*, 256–257.
4. Fujikoshi, Y. (2020). Computable error bounds for asymptotic approximations of the quadratic discriminant function when the sample sizes and dimensionality are large. *Hiroshima Mathematical Journal, 50*.
5. Fujikoshi, Y., & Seo, T. (1998). Asymptotic approximations for EPMC's of the linear and the quadratic discriminant functions when the samples sizes and the dimension are large. *Statistics Analysis Random Arrays, 6*, 269–280.
6. Fujikoshi, Y., Ulyanov, V. V., & Shimizu, R. (2010). *Multivariate analysis: High-dimensional and large-sample approximations*. Hoboken: Wiley.
7. Lachenbruch, P. A. (1968). On expected probabilities of misclassification in discriminant analysis, necessary sample size, and a relation with the multiple correlation coefficients. *Biometrics, 24*, 823–834.
8. Okamoto, M. (1963). An asymptotic expansion for the distribution of the linear discriminant function. *The Annals of Mathematical Statistics, 34*, 1286–1301.
9. Raudys, S. (1972). On the amount of priori information in designing the classification algorithm. *Technical. Cybernetics, 4*, 168–174. (in Russian).
10. Shimizu, R., & Fujikoshi, Y. (1997). Sharp error bounds for asymptotic expansions of the distribution functions of scale mixtures. *Annals of the Institute of Statistical Mathematics, 49*, 285–297.
11. Siotani, M. (1982). Large sample approximations and asymptotic expansions of classification statistic. In P. R. Krishnaiah & L. N. Kanal (Eds.), *Handbook of statistics* (Vol. 2, pp. 47–60). Amsterdam: North-Holland Publishing Company.

12. Yamada, T., Sakurai, T., & Fujikoshi, Y. (2017). High-dimensional asymptotic results for EPMCs of W- and Z-rules. *Hiroshima Statistical Research Group, TR*, *17*, 12.
13. Wyman, F. J., Young, D. M., & Turner, D. W. (1990). A comparison of asymptotic error rate expansions for the sample linear discriminant function. *Pattern Recognition*, *23*, 775–783.

Chapter 5
Cornish–Fisher Expansions

Abstract First, we focus on Cornish–Fisher expansions for quantiles whose distributions have Edgeworth-type expansions. Then, when the Edgeworth-type expansions have computable error bounds, we give computable error bounds for the Cornish–Fisher expansions. Some of the results depend on Bartlett-type corrections. The results are illustrated by examples.

5.1 Introduction

Let X be a univariate random variable with a continuous distribution function F. For $\alpha : 0 < \alpha < 1$, there exists x such that $F(x) = \alpha$, which is called the (lower) $100\alpha\%$ *point* of F. If F is strictly increasing, the inverse function $F^{-1}(\cdot)$ is well defined and the $100\alpha\%$ point is uniquely determined. We refer to "quantiles" without reference to particular values of α to mean the values given by $F^{-1}(\cdot)$. Even in the general case, when $F(x)$ is not necessarily continuous nor is it strictly increasing, we can define its inverse function by formula

$$F^{-1}(u) = \inf\{x;\ F(x) > u\}.$$

This is a right-continuous non-decreasing function defined on the interval $(0, 1)$ and $F(x_0) \geq u_0$ if $x_0 = F^{-1}(u_0)$.

In this chapter we are interested in asymptotic expansion of the quantile of $F(x)$ under the assumption that $F(x)$ has an asymptotic expansion. Such an expansion is called a Cornish–Fisher expansion. We provide error bounds for asymptotic expansions of the quantile when $F(x)$ has an asymptotic expansion with computable error bounds. Such results are given when the error bounds are $\mathrm{O}(\epsilon)$ and $\mathrm{O}(\epsilon^2)$, where $\epsilon = n^{-1/2}$ or $\epsilon = n^{-1}$ with sample size n. The results when the error bounds are $\mathrm{O}(\epsilon^2)$ depend on a Bartlet-type correction. The results are illustrated by examples.

Y. Fujikoshi and V. V. Ulyanov, *Non-Asymptotic Analysis of Approximations for Multivariate Statistics*, JSS Research Series in Statistics,
https://doi.org/10.1007/978-981-13-2616-5_5

5.2 Cornish–Fisher Expansion

In this section, we explore asymptotic approximations of the quantiles of a given distribution function $F_n(x)$ with a parameter $n = 1, 2, \ldots$. We assume that each F_n admits the Edgeworth-type expansion (EE) in the powers of $\epsilon = n^{-1/2}$ or n^{-1}:

$$\begin{aligned} F_n(x) &= G_{k,n}(x) + R_k(x) \quad \text{with } R_k(x) = O(\epsilon^k) \quad \text{and} \\ G_{k,n}(x) &= G(x) + \left\{\epsilon a_1(x) + \cdots + \epsilon^{k-1} a_{k-1}(x)\right\} g(x), \end{aligned} \tag{5.1}$$

where $g(x)$ is the density function of the limiting distribution function $G(x)$. Then we can approximate quantiles of F_n by the corresponding quantiles of $G_{k,n}$. An important approach to the problem of approximating the quantiles of F_n is to use their asymptotic relation to quantiles of G. Let x and u be the corresponding quantiles of F_n and G, respectively. Then we have

$$F_n(x) = G(u). \tag{5.2}$$

Let $x(u)$ and $u(x)$ denote the solutions of (5.2) for x in terms of u and u in terms of x, respectively, i.e., $u(x) = G^{-1}(F_n(x))$ and $x(u) = F_n^{-1}(G(u))$. Then we can use the EE (5.1) to obtain formal solutions $x(u)$ and $u(x)$ in the form

$$x(u) = u + \epsilon b_1(u) + \epsilon^2 b_2(u) + \cdots \tag{5.3}$$

and

$$u(x) = x + \epsilon c_1(x) + \epsilon^2 c_2(x) + \cdots. \tag{5.4}$$

The first few terms of these expansions were obtained by [3, 6] when G is the standard normal distribution function (i.e., $G = \Phi$). Both (5.3) and (5.4) are called Cornish–Fisher expansions (CFE). Usually a CFE is applied in the following form with $k = 1$, 2 or 3:

$$x_k(u) = u + \sum_{j=1}^{k-1} \epsilon^j b_j(u) + \hat{R}_k(u) \quad \text{with } \hat{R}_k(u) = O(\epsilon^k). \tag{5.5}$$

It is known (see, e.g., [18]) how to find explicit expressions for $b_1(u)$ and $b_2(u)$ as soon as we have (5.1). By Taylor expansions for G, g, and a_1, we obtain

$$\begin{aligned} b_1 &= -a_1(u), \\ b_2 &= \frac{1}{2}\{g'(u)/g(u)\}a_1^2(u) - a_2(u) + a_1'(u)a_1(u), \end{aligned} \tag{5.6}$$

provided that g, a_1, and a_2 are sufficiently smooth functions. Now we seek for explicit expressions for $b_1(u)$ and $b_2(u)$. Substituting (5.3) or (5.5) with $k = 3$ to (5.2) and using (5.1), we have

$$\begin{aligned}F_n(x) &= F_n(u + \epsilon b_1 + \epsilon^2 b_2 + \cdots)\\ &= G(u + \epsilon b_1 + \epsilon^2 b_2) + g(u + \epsilon b_1 + \epsilon^2 b_2)\\ &\quad \times \{\epsilon a_1(u + \epsilon b_1) + \epsilon^2 a_2(u)\} + \mathrm{O}(\epsilon^3).\end{aligned}$$

Considering Taylor expansions for G, g, and a_1, we obtain

$$\begin{aligned}F_n(x) = G(u) &+ \epsilon g(u)\{b_1 + a_1(u)\}\\ &+ \epsilon^2\Big[g(u)b_2 + \frac{1}{2}g'(u)b_1^2 + g(u)a_1'(u)b_1\\ &\qquad + g(u)a_2(u) + g'(u)b_1a_1(u)\Big] + \mathrm{O}(\epsilon^3),\end{aligned}$$

which should be $G(u)$. Therefore,

$$\begin{aligned}b_1 &= -a_1(u),\\ b_2 &= \frac{1}{2}\{g'(u)/g(u)\}a_1^2(u) - a_2(u) + a_1'(u)a_1(u).\end{aligned}$$

These results are generalized as the generalized Cornish–Fisher expansion formula in the following theorem.

Theorem 5.1 *Assume that the distribution function G is analytic and the density function $g(x)$ and G are arbitrarily differentiable. Then the following relations for x and u satisfying $F_n(x) = G(u)$ hold:*

$$x = u - \sum_{r=1}^{\infty} \frac{1}{r!}\Big\{-[g(u)]^{-1}d_u\Big\}^{r-1}\Big[\{z_n(u)\}^r/g(u)\Big], \tag{5.7}$$

where $d_u \equiv d/du$ and $z_n(u) = F_n(u) - G(u)$.

Proof See [16]. □

In many statistical applications, $F_n(x)$ is known to have an asymptotic expansion of the form

$$F_n(x) = G(x) + g(x)\Big[n^{-a}p_1(x) + n^{-2a}p_2(x) + \cdots\Big], \tag{5.8}$$

where $p_r(x)$ may be polynomials in x and $a = 1/2$ or 1. Then formula (5.7) can be written as

$$x = u - \sum_{r=1}^{\infty}\frac{1}{r!}d_{(r)}\{q_n(u)\}^r, \tag{5.9}$$

where $q_n(u) = n^{-a}p_1(u) + n^{-2a}p_2(u) + \cdots$, $m(x) = -g'(x)/g(x)$,

$$d_{(1)} = \text{the identity operator,}$$
$$d_{(r)} = \{m(u) - d_u\}\{2m(u) - d_u\} \cdots \{(r-1)m(u) - d_u\}, \quad r = 2, 3, \ldots.$$

The rth term in (5.9) is $O(n^{-ra})$.

5.3 Error Bounds for Cornish–Fisher Expansion

In the following theorems we show how $x_k(u)$ from (5.5) could be expressed in terms of u. Moreover, we show what kind of bounds we can obtain for $\hat{R}_k(x)$ as soon as we have bounds for $R_k(x)$ from (5.1), based on [23].

Theorem 5.2 *Suppose that for the distribution function of a statistic U we have*

$$F(x) \equiv \Pr\{U \leq x\} = G(x) + R_1(x), \tag{5.10}$$

where for remainder term $R_1(x)$ there exists a constant c_1 such that

$$|R_1(x)| \leq d_1 \equiv c_1\epsilon.$$

Let x_α and u_α be the upper $100\alpha\%$ points of F and G, respectively, that is

$$\Pr\{U \leq x_\alpha\} = G(u_\alpha) = 1 - \alpha. \tag{5.11}$$

Then for any α such that $1 - c_1\epsilon > \alpha > c_1\epsilon > 0$ we have

(i) $u_{\alpha+d_1} \leq x_\alpha \leq u_{\alpha-d_1}$.
(ii) $|x_\alpha - u_\alpha| \leq c_1\epsilon / g(u_{(1)})$, *where g is the density function of the limiting distribution G and*

$$g(u_{(1)}) = \min_{u \in [u_{\alpha+d_1}, u_{\alpha-d_1}]} g(u).$$

Theorem 5.3 *In terms of the notation of Theorem 5.2 we assume that*

$$F(x) \equiv \Pr\{U \leq x\} = G(x) + \epsilon g(x)a(x) + R_2(x),$$

where for remainder term $R_2(x)$ there exists a constant c_2 such that

$$|R_2(x)| \leq d_2 \equiv c_2\epsilon^2.$$

Let $T = T(u)$ be a monotone increasing transform such that

$$\Pr\{T(U) \leq x\} = G(x) + \tilde{R}_2(x) \quad \textit{with} \quad |\tilde{R}_2(x)| \leq \tilde{d}_2 \equiv \tilde{c}_2\epsilon^2.$$

Let $\tilde{x}_\alpha$ *and* u_α *be the upper* $100\alpha\%$ *points of* $\Pr\{T(U) \le x\}$ *and* G, *respectively. Then for any* α *such that*

$$1 - \tilde{c}_2\epsilon^2 > \alpha > \tilde{c}_2\epsilon^2 > 0,$$

it holds that

$$|\tilde{x}_\alpha - u_\alpha| \le \tilde{c}_2\epsilon^2 / g(u_{(2)}), \tag{5.12}$$

where

$$g(u_{(2)}) = \min_{u \in [u_{\alpha+\tilde{d}_2}, u_{\alpha-\tilde{d}_2}]} g(u).$$

Theorem 5.4 *We use the notation of Theorem 5.3. Let* $b(x)$ *be a function inverse to* T, *i.e.,* $b(T(x)) = x$. *Then* $x_\alpha = b(\tilde{x}_\alpha)$ *and for* α *such that* $1 - \tilde{c}_2\epsilon^2 > \alpha > \tilde{c}_2\epsilon^2$ *we have*

$$|x_\alpha - b(u_\alpha)| \le \tilde{c}_2 \frac{|b'(u^*)|}{g(u_{(2)})} \epsilon^2, \tag{5.13}$$

where

$$|b'(u^*)| = \max_{u \in [u_{\alpha+\tilde{d}_2}, u_{\alpha-\tilde{d}_2}]} |b'(u)|.$$

Moreover,

$$b(x) = x - \epsilon a(x) + O(\epsilon^2). \tag{5.14}$$

Remark 5.1 The main assumption of the theorems is that for distributions of statistics and for distributions of transformed statistics we have some approximations with computable error bounds. There is not much literature with these kinds of non-asymptotic results, because generating such results requires a technique which is different from prevailing asymptotic methods (cf., e.g., [10, 24]). In a series of papers [1, 8, 9, 11, 12, 20–22], we derived non-asymptotic results for a wide class of statistics including multivariate scale mixtures and MANOVA tests, in the case of high dimensions, that is the case when the dimension of observations and sample size are comparable. Some of the results were included in the book [10].

Remark 5.2 The results of Theorems 5.2–5.4 cannot be extended to the whole range of $\alpha \in (0, 1)$. This follows from the fact that the Cornish–Fisher expansion does not converge uniformly in $0 < \alpha < 1$. See the corresponding example in Sect. 2.5 of [14].

Remark 5.3 In Theorem 5.4, we required the existence of a monotone increasing transform $T(z)$ such that distribution of transformed statistic $T(U)$ is approximated by some limit distribution $G(x)$ in a better way than the distribution of the original statistic U. We call this transformation $T(z)$ the Bartlett-type correction.

Remark 5.4 According to (5.13) and (5.14) the function $b(u_\alpha)$ in Theorem 5.4 could be considered as an "asymptotic expansion" for x_α up to order $O(\epsilon^2)$.

5.4 Proofs for Error Bounds

Proof of Theorem 5.2 By the mean value theorem,

$$|G(x_\alpha) - G(u_\alpha)| \geq |x_\alpha - u_\alpha| \min_{0<\theta<1} g(u_\alpha + \theta(x_\alpha - u_\alpha)).$$

From (5.10) and the definition of x_α and u_α in (5.11), we have

$$\begin{aligned} |G(x_\alpha) - G(u_\alpha)| &= |G(x_\alpha) - \Pr\{U \leq x_\alpha\}| \\ &= |R_1(x_\alpha)| \leq d_1. \end{aligned}$$

Therefore,

$$|x_\alpha - u_\alpha| \leq \frac{d_1}{\min_{0<\theta<1} g(u_\alpha + \theta(x_\alpha - u_\alpha)}. \tag{5.15}$$

On the other hand, it follows from (5.10) that

$$\begin{aligned} G(x_\alpha) &= 1 - \alpha - R_1(\alpha) \\ &\leq 1 - (\alpha - d_1) = G(u_{\alpha - d_1}). \end{aligned}$$

This implies that $x_\alpha \leq u_{\alpha-d_1}$. Similarly, we have $u_{\alpha+d_1} \leq x_\alpha$. Therefore, Theorem 5.2 (i) is proved. It follows from Theorem 5.2(i) that

$$\min_{u \in [u_{\alpha+d_1}, u_{\alpha-d_1}]} g(u) \leq \min_{0<\theta<1} g(u_\alpha + \theta(x_\alpha - u_\alpha)).$$

Thus, using (5.15) we get the statement of Theorem 5.2(ii).

Proof of Theorem 5.3 It is straightforward to identify that it is sufficient to apply Theorem 5.1(ii) to the transformed statistic $T(U)$.

Proof of Theorem 5.4 Using (5.12) and the mean value theorem we obtain

$$\tilde{x}_\alpha - u_\alpha = b^{-1}(x_\alpha) - b^{-1}(b(u_\alpha)) = (b^{-1})'(x^*)\big(x_\alpha - b(u_\alpha)\big), \tag{5.16}$$

where x^* is a point on the interval $\big(\min\{x_\alpha, b(u_\alpha)\}, \max\{x_\alpha, b(u_\alpha)\}\big)$.

By Theorem 5.2(i) we have

$$u_{\alpha+\tilde{d}_2} \leq \tilde{x}_\alpha \leq u_{\alpha-\tilde{d}_2}.$$

Therefore, for $x_\alpha = b(\tilde{x}_\alpha)$ we get

$$\big(\min\{b^{-1}(x_\alpha), u_\alpha\}, \max\{b^{-1}(x_\alpha), u_\alpha\}\big) \subseteq \big(u_{\alpha+\tilde{d}_2}, u_{\alpha-\tilde{d}_2}\big). \tag{5.17}$$

Since by properties of derivatives of inverse functions

$$(b^{-1})'(z) = 1/b'(b^{-1}(z)) = 1/b'(y)$$

for $z = b(y)$, the relations (5.16) and (5.17) imply (5.13). Representation (5.14) for $b(x)$ follows from (5.6) and (5.13).

5.5 Transformations for Improved Approximations

Suppose that a statistic T has an asymptotic normal or chi-square approximation as some parameter; typically, the sample size n tends to infinity. In this case, it is interesting to construct improved normal or chi-square approximations for T. We consider this problem by considering a transformation $\widetilde{T} = \widetilde{T}(T)$, especially such that $\widetilde{T}(T)$ is monotone. The monotonicity of a transformation is important in keeping optimality of a statistic T and computing the probability $\Pr\{T \ge x\}$.

First, consider the case in which the limiting distribution of T as n tends to infinity is normal. Suppose that T has an asymptotic expansion

$$\Pr\{T \le x\} = \Phi(x) + \frac{1}{\sqrt{n}}\phi(x)(ax^2 + b) + \mathrm{o}(n^{-1/2}), \tag{5.18}$$

which is a typical form. In general, we are interested in a transformation that leads to the disappearance of the terms of $\mathrm{O}(n^{-1/2})$ in the asymptotic expansion. It is seen that such a transformation is given by $T^* = T + (aT^2 + b)/\sqrt{n}$. However, this transformation is not monotone. As a method of making T^* monotone, we may consider

$$\widetilde{T} = T^* + \frac{1}{3n}a^2T^3. \tag{5.19}$$

The transformation $\widetilde{T}$ is monotone, since $\widetilde{T}' = (1 + aT)^2 > 0$. Further, the added term is $\mathrm{O}(n^{-1})$. Therefore,

$$\Pr\{\widetilde{T} \le x\} = \Phi(x) + \mathrm{o}(n^{-1}).$$

This idea was used in [15]. This transformation removes its skewness.

Next we consider the case when T is asymptotically distributed to χ^2_q. There is a wide class of statistics T allowing the following representation:

$$\Pr\{T \le x\} = G_q(x) + \frac{1}{n} \sum_{j=0}^{k} a_j G_{q+2j}(x) + R_{2k}, \tag{5.20}$$

where $R_{2k} = O(n^{-2})$ and $G_q(x)$ is the distribution function of a chi-squared distribution with q degrees of freedom and coefficients a_j satisfy the relation $\sum_{j=0}^{k} a_j = 0$. Examples of T in MANOVA tests are as follows: for $k = 1$, the likelihood ratio test statistic; for $k = 2$, the Lawley–Hotelling criterion and the Bartlett–Nanda–Pillai criterion, which are test statistics for multivariate linear hypotheses under normality; for $k = 3$, the score test statistic and Hotelling's T^2-statistic under nonnormality. For such statistics, sufficient conditions have been studied for a transformation $\widetilde{T}(T)$ to be a Bartlett-type correction (see Remark 5.3 above). A well-known Bartlett-type correction is given by

$$\widetilde{T} = \left(1 - \frac{1}{n} b_1\right) T, \tag{5.21}$$

where $b_1 = (2/q) \sum_{j=1}^{k} j a_j$. Then, $\mathrm{E}(\widetilde{T}) = q + O(n^{-2})$. So, $\widetilde{T}$ has a better chi-square approximation.

As a more improved chi-square approximation, there are some transformations satisfying (1) $\Pr(\widetilde{T} \le x) = G_q(x) + O(n^{-2})$, and (2) $\widetilde{T}$ is monotone. Some of the results are given in [4, 5, 7, 13, 17, 19]. Let us consider a polynomial transformation under which the terms of $O(n^{-1})$ in its asymptotic expansion disappear is given (e.g., [2]) by

$$T^* = T \left(1 - \frac{1}{n} \sum_{j=1}^{k} c_j T^{j-1}\right),$$

where approximation $c_j = 2 \sum_{\ell=j}^{k} a_\ell / \prod_{\ell=j}^{j-1} (f + 2\ell)$. This polynomial transformation is not monotone. However, it is known (see, e.g., [17]) that by modifying T^* we have a monotone transformation given by

$$\widetilde{T} = T^* + \frac{1}{4n^2} \sum_{i,j=1}^{k} \frac{c_i c_j i j}{i + j - 1} T^{i+j-1}.$$

It may be noted that as in the abovementioned results, there are many statistics having an improved chi-square approximation. However, the number of papers on explicit error bounds is relatively small compared to those on improved chi-square approximations. We note that Theorem 4.1 of [19] gives explicit error bounds for an improved chi-square approximation of T in (5.20).

5.6 Examples

Here, first we consider the null distribution of Hotelling's generalized T_0^2 (or Lawley and Hotelling criterion) statistic defined by $T_0^2 = n\,\mathrm{tr}\mathbf{S}_h\,\mathbf{S}_e^{-1}$, where $\mathbf{S}_h$ and $\mathbf{S}_e$ are independently distributed as Wishart distributions $\mathrm{W}_p(q, \mathbf{I}_p)$ and $\mathrm{W}_p(n, \mathbf{I}_p)$, respectively. In Theorem 4.1(ii) in [12], we have the following computable error bound for $n \geq p$: Let $T = T_0^2$.

$$|\Pr(T \leq x) - G_r(x) - \frac{r}{4n}\{(q-p-1)G_r(x)$$
$$-2qG_{r+2}(x) + (q+p+1)G_{r+4}(x)\}| \leq \frac{c_{p,q}}{n^2},$$

where $r = pq$ and for constant $c_{p,q}$ we gave an expicit formula with dependence on p and q. Therefore, according to [19], we can use the Bartlett-type correction $\widetilde{T}(t)$ as

$$T(z) = \frac{a-1}{2b} + \sqrt{\left(\frac{a-1}{2b}\right)^2 + \frac{t}{b}},$$

where

$$a = \frac{1}{2n}p(q-p-1), \quad b = \frac{1}{2n}p(q+p+1)(q+2) - 1.$$

It is clear that $\widetilde{T}(t)$ is invertible and so we can apply Theorem 5.3 with the help of Theorem 4.1 in [19].

One more example is connected with the sample correlation coefficient. Let (X_j, Y_j), $j = 1, \ldots, n$ be a sample of size n from a 2-dimensional normal population $\mathrm{N}(\mathbf{0}, \mathbf{I}_2)$. Consider the sample correlation coefficient

$$R = \frac{\sum_{j=1}^n X_j Y_j}{\sqrt{\sum_{j=1}^n X_j^2 \sum_{j=1}^n Y_j^2}}.$$

Let $T = \sqrt{N}R$. In [2] it was proved that for $n \geq 7$ and $N = n - 2.5$,

$$\sup_x \left|\Pr(T \leq x) - \Phi(x) - \frac{1}{4}x^3\varphi(x)N^{-1}\right| \leq B_n N^{-2},$$

with $B_n \leq 2.2$. It is straightforward to see that we can take $\widetilde{T}(t)$ as the Bartlett-type correction in the form $\widetilde{T}(t) = t + t^3/(4N)$. Then the inverse function $b(t) = \widetilde{T}^{-1}(t)$ is defined by

$$\begin{aligned} b(t) &= \left(2Nz + \sqrt{(2Nz)^2 + (4N/3)^3}\right)^{1/3} \\ &\quad - \left(-2z + \sqrt{(2Nz)^2 + (4N/3)^3}\right)^{1/3} \\ &= z - \frac{z^3}{4N} + \frac{3z^5}{16N^2} + \mathrm{O}(N^{-3}). \end{aligned}$$

Now we can apply Theorem 5.3, by deriving non-asymptotic bound for $\widetilde{T}(t)$ in a smilar way as in [1].

References

1. Christoph, G., Ulyanov, V. V., & Fujikoshi, Y. (2013). Accurate approximation of correlation coefficients by short Edgeworth–Chebyshev expansion and its statistical applications. *Springer Proceedings in Mathematics and Statistics*, *33*, 239–260.
2. Cordeiro, G. M., & de Paula Ferrari, S. L. (1991). A modified score test statistic having chi-squared distribution to order n^{-1}. *Biometrika*, *78*, 573–582.
3. Cornish, E. A., & Fisher, R. A. (1937). Moments and cumulants in the specification of distributions. *Revue de l'Institut international de Statistique*, *4*, 307–320.
4. Enoki, H., & Aoshima, M. (2004). Transformations with improved chi-squared approximations. In *Proceedings of Symposium, Research Institute for Mathematical Science, Kyoto University* (Vol. 1380, pp. 160–181).
5. Enoki, H., & Aoshima, M. (2006). Transformations with improved asymptotic approximations and their accuracy. *SUT Journal of Mathematics*, *42*(1), 97–122.
6. Fisher, R. A., & Cornish, E. A. (1946). The percentile points of distributions having known cumulants. *Journal of American Statistical Association*, *80*, 915–922.
7. Fujikoshi, Y. (2000). Transformations with improved chi-squared approximations. *Journal of Multivariate Analysis*, *73*, 249–263.
8. Fujikoshi, Y., & Ulyanov, V. V. (2006). Error bounds for asymptotic expansions of Wilks' lambda distribution. *Journal of Multivariate Analysis*, *97*, 1941–1957.
9. Fujikoshi, Y., & Ulyanov, V. V. (2006). On accuracy of approximations for location and scale mixtures. *Journal of Mathematical Sciences*, *138*, 5390–5395.
10. Fujikoshi, Y., Ulyanov, V. V., & Shimizu, R. (2010). *Multivariate statistics: High-dimensional and large-sample approximations*. Wiley series in probability and statistics. Hoboken: Wiley.
11. Fujikoshi, Y., Ulyanov, V. V., & Shimizu, R. (2005). L_1-norm error bounds for asymptotic expansions of multivariate scale mixtures and their applications to Hotelling's generalized T_0^2. *Journal of Multivariate Analysis*, *96*, 1–19.
12. Fujikoshi, Y., Ulyanov, V. V., & Shimizu, R. (2005). Error bounds for asymptotic expansions of the distribution of multivariate scale mixture. *Hiroshima Mathematical Journal*, *35*, 453–469.
13. Fujisawa, H. (1997). Improvement on chi-squared approximation by monotone transformation. *Journal of Multivariate Analysis*, *60*, 84–89.
14. Hall, P. (1992). *The Bootstrap and Edgeworth expansion*. New York: Springer.
15. Hall, P. (1992). On the removal of skewness by transformation. *Journal of the Royal Statistical Society: Series B*, *54*, 221–228.
16. Hill, G. W., & Davis, A. W. (1968). Generalized asymptotic expansions of Cornish–Fisher type. *The Annals of Mathematical Statistics*, *39*, 1264–1273.
17. Kakizawa, Y. (1996). Higher order monotone Bartlett-type adjustment for some multivariate test statistics. *Biometrika*, *83*, 923–927.
18. Ulyanov, V. V. (2011). Cornish–Fisher expansions. In M. Lovric (Ed.), *International encyclopedia of statistical science* (pp. 312–315). Berlin: Springer.

19. Ulyanov, V. V., & Fujikoshi, Y. (2001). On accuracy of improved χ^2-approximations. *Georgian Mathematical Journal, 8*(2), 401–414.
20. Ulyanov, V. V., Fujikoshi, Y., & Shimizu, R. (1999). Nonuniform error bounds in asymptotic expansions for scale mixtures under mild moment conditions. *Journal of Mathematical Sciences, 93*, 600–608.
21. Ulyanov, V. V., Christoph, G., & Fujikoshi, Y. (2006). On approximations of transformed chi-squared distributions in statistical applications. *Siberian Mathematical Journal, 47*, 1154–1166.
22. Ulyanov, V. V., Wakaki, H., & Fujikoshi, Y. (2006). Berry–Esseen bound for high dimensional asymptotic approximation of Wilks' lambda distribution. *Statistics and Probability Letters, 76*, 1191–1200.
23. Ulyanov, V. V., Aoshima, M., & Fujikoshi, Y. (2016). Non-asymptotic results for Cornish-Fisher expansions. *Journal of Mathematical Statistics, 218*, 363–368.
24. Wakaki, H., Fujikoshi, Y., & Ulyanov, V. V. (2014). Asymptotic expansions of the distributions of MANOVA test statistics when the dimension is large. *Hiroshima Mathematical Journal, 44*, 247–259.

Chapter 6
Likelihood Ratio Tests with Box-Type Moments

Abstract In this chapter we consider statistics with Box-types of moments. Such statistics appear as various likelihood ratio statistics for multivariate normal populations. First, their large-sample approximation method is explained. Then, we derive their high-dimensional asymptotic expansions. Further, it is noted that an error bound for high-dimensional asymptotic expansions can be derived for some statistics including the lambda distribution.

6.1 Introduction

Related to the distributions of likelihood ratio criteria in multivariate normal models, consider a general statistic $W (0 \le W \le 1)$ which has moments of the form

$$\mathrm{E}(W^h) = \left(\frac{\prod_{k=1}^{b} y_k^{y_k}}{\prod_{j=1}^{a} x_j^{x_j}} \right)^h \frac{\prod_{j=1}^{a} \Gamma[x_j(1+h) + \xi_j] \prod_{k=1}^{b} \Gamma[y_k + \eta_k]}{\prod_{j=1}^{a} \Gamma[x_j + \xi_j] \prod_{k=1}^{b} \Gamma[y_k(1+h) + \eta_k]}, \tag{6.1}$$

where $\sum_{j=1}^{a} x_j = \sum_{k=1}^{b} y_k$. The moment expressed in the form (6.1) is called Box-type moment. Distributional results under a large-sample framework can be found in [6, Chap. 10], [2, Chap. 8], [7, Chap. 6] and [5, Chap. 5].

As a special statistic with Box-type of moments, consider the lambda distribution defined by

$$\Lambda = \frac{|\mathbf{S}_e|}{|\mathbf{S}_e + \mathbf{S}_h|} \sim \Lambda_p(q, n), \tag{6.2}$$

where $\mathbf{S}_h$ and $\mathbf{S}_e$ are independently distributed as $\mathrm{W}_p(q, \boldsymbol{\Sigma})$ and $\mathrm{W}_p(n, \boldsymbol{\Sigma})$, respectively. The Λ appears, for example, as a likelihood ratio test for testing the equality of mean vectors $\boldsymbol{\mu}_i$, $i = 1, \ldots, q+1$, based on an N_i sample from $\mathrm{N}_p(\boldsymbol{\mu}, \boldsymbol{\Sigma})$. In this case, $n = N - (q+1)$ and $N = N_1 + \cdots + N_{q+1}$. When we consider the distribution of Λ, we may assume $\boldsymbol{\Sigma} = \mathbf{I}_p$. The likelihood ratio criterion is based on $\lambda = \Lambda^{n/2}$. The hth moment of Λ is given by

Y. Fujikoshi and V. V. Ulyanov, *Non-Asymptotic Analysis of Approximations for Multivariate Statistics*, JSS Research Series in Statistics,
https://doi.org/10.1007/978-981-13-2616-5_6

$$\mathrm{E}[\Lambda^h] = \prod_{j=1}^{p} \frac{\Gamma[\frac{1}{2}(n-j+1)+h]\Gamma[\frac{1}{2}(n+q-j+1)]}{\Gamma[\frac{1}{2}(n-j+1)]\Gamma[\frac{1}{2}(n+q-j+1)+h]}. \tag{6.3}$$

In Sect. 6.2 we explain a well-known asymptotic expansion method by considering $-2\log\Lambda^{n/2}$ in stead of $-2\log W$, when the sample is large. In general, large-sample approximations worsen when the dimensions are large. As an approach to covering the weak point, an asymptotic expansion of the distribution of $-\log\Lambda$ was given by [8] under a high-dimensional asymptotic framework;

$$q;\ \text{fixed},\ p \to \infty,\ n \to \infty,\ p/n \to c \in (0,1). \tag{6.4}$$

In Sect. 6.3 we consider high-dimensional approximations where the sample size and dimensions are both large. In Sect. 6.4 we give a method of obtaining their error bounds which are applicable for some statistics including the lambda distribution.

6.2 Large-Sample Asymptotic Expansions

For simplicity, consider an asymptotic expansion of $T = -2\log\Lambda^{n/2} = -n\log\Lambda$ under a large-sample framework;

$$p, q;\ \text{fixed},\ n \to \infty. \tag{6.5}$$

The characteristic function of T is expressed as

$$\begin{aligned} C(t) &= \mathrm{E}\left[e^{it(-n\log\Lambda)}\right] \\ &= \mathrm{E}\left[\Lambda^{-nit}\right] \\ &= \prod_{j=1}^{p} \frac{\Gamma[\frac{1}{2}n(1-2it)+\frac{1}{2}(-j+1)]\Gamma[\frac{1}{2}n+\frac{1}{2}(q-j+1)]}{\Gamma[\frac{1}{2}n+\frac{1}{2}(-j+1)]\Gamma[\frac{1}{2}n(1-2it)+\frac{1}{2}(q-j+1)]}. \end{aligned} \tag{6.6}$$

We can derive an expansion for $C(t)$ by using the following generalized version of Stirling's formula for the gamma function

$$\begin{aligned} \log\Gamma(z+h) &= \log\sqrt{2\pi} + \left(z+h-\frac{1}{2}\right)\log z - z \\ &\quad + \frac{1}{2z}\left(h^2 - h + \frac{1}{6}\right) + \mathrm{O}(z^{-2}). \end{aligned}$$

For example, we can derive

$$C(t) = (1 - 2it)^{-pq/2}\left[1 + \frac{pq}{4n}(q - p - 1)\left\{1 - (1 - 2it)^{-1}\right\} + \mathrm{O}(n^{-2})\right].$$

Inverting the above characteristic function formally, we can derive a formal asymptotic expansion,

$$\begin{aligned} \mathrm{P}(-n \log \Lambda \le x) &= G_f(x) \\ &\quad + \frac{pq}{4n}(q - p - 1)\left\{G_f(x) - G_{f+2}(x)\right\} + \mathrm{O}(n^{-2}), \end{aligned} \tag{6.7}$$

where $f = pq$ and $G_f(x)$ is the distribution function of χ^2_f.

The above derivation method is very powerful in finding asymptotic expansions. For an asymptotic expansion of $-2 \log W$, see [2, Chap. 8]. However, it is not clear whether we can extend it useful finding an error bound. On the other hand, a computable error bound for the above expansion formula (6.7) can be derived, based on an error bound in L_1-norm for a multivariate scale mixture, see Chaps. 3 and [4].

6.3 High-Dimensional Asymptotic Expansions

In general, large-sample approximations as in (6.7) worsen when the dimensions are large. As an approach to covering the weak point, we consider high-dimensional approximations where the sample size and dimensions are both large. An asymptotic expansion of the distribution of $-\log \Lambda$ under (6.4) was obtained by using that

$$\Lambda_p(q, n) = \Lambda_q(p, n - p + q), \tag{6.8}$$

and $\Lambda_q(p, n - p + q)$ is the same distribution of $\prod_{j=1}^{q} X_j$, where $X_1, \ldots, X_q$ are independent and X_j is distributed as beta distribution with degrees of freedom $p/2$ and $(n - p + j)/2$. Using this fact, the hth moment of Λ is also expressed as

$$\mathrm{E}[\Lambda^h] = \prod_{j=1}^{q} \frac{\Gamma[\frac{1}{2}(n - p + j) + h]\Gamma[\frac{1}{2}(n + j)]}{\Gamma[\frac{1}{2}(n - p + j)]\Gamma[\frac{1}{2}(n + j) + h]}. \tag{6.9}$$

We consider asymptotic expansions of W with moments (6.1) under the assumption that the sample size and the dimension are large. In most applications,

$$\mathrm{C1}: \ x_j = c_j m, \ \ i = 1, \ldots, a; \quad y_k = d_k m, \ \ k = 1, \ldots, b, \tag{6.10}$$

where c_i and d_j are positive constants such that $\sum_{j=1}^{a} c_j = \sum_{k=1}^{b} d_k$, and m is a sample size or a number growing with the sample size. Let $\widetilde{C}(t)$ be the characteristic function of $\widetilde{T} = -(2/m) \log W$. Then, the cumulant generating function of $\widetilde{T}$ can be expressed as

$$\begin{aligned}\log \widetilde{C}(t) &= \log \mathrm{E}[\exp(it\widetilde{T})] = \log \mathrm{E}[W^{-2it/m}] \\ &= -2it\left\{\sum_{j=1}^{a} c_j \log c_j - \sum_{k=1}^{b} d_k \log d_k\right\} \\ &+ \sum_{j=1}^{a}\{\log \Gamma[c_j m + \xi_j - 2c_j it] - \log \Gamma[c_j m + \xi_j]\} \\ &+ \sum_{k=1}^{b}\{\log \Gamma[d_k m + \eta_k - 2d_k it] - \log \Gamma[d_k m + \eta_k]\}.\end{aligned}$$

Now we use the Taylor expansion formula:

$$\log \Gamma(a+b) = \log \Gamma(a) + \sum_{k=1}^{\infty} \frac{1}{k!}\psi^{(k-1)}(a) b^k,$$

where ψ is the digamma function defined by $\psi(z) = (d/dz)\log \Gamma(z)$. It is known that the polygamma function $\psi^{(s)}(a) = d^s\psi(z)/dz^s|_{z=a}$ can be expressed as

$$\psi^{(s)}(a) = \begin{cases} -C + \displaystyle\sum_{k=0}^{\infty}\left(\frac{1}{1+k} - \frac{1}{k+a}\right), & s = 0, \\ \displaystyle\sum_{k=0}^{\infty} \frac{(-1)^{s+1}s!}{(k+a)^{s+1}}, & s = 1, 2, \ldots, \end{cases}$$

where C is the Euler constant. Using the above Taylor formula, we obtain

$$\log \widetilde{C}(t) = \sum_{s=1}^{\infty} \frac{\kappa^{(s)}}{s!}(it)^s, \tag{6.11}$$

where

$$\begin{aligned}\kappa^{(1)} &= -2\left\{\sum_{j=1}^{a} c_j \log c_j - \sum_{k=1}^{b} d_k \log d_k\right\} \\ &- \left\{\sum_{j=1}^{a} 2c_j\psi^{(0)}(c_j m + \xi_j) - \sum_{k=1}^{b} 2d_k\psi^{(0)}(d_k m + \eta_k)\right\}\end{aligned}$$

and for $s \geq 2$

$$\kappa^{(s)} = (-1)^s \left\{ \sum_{j=1}^{a} (2c_j)^s \psi^{(s-1)}(c_j m + \xi_j) - \sum_{k=1}^{b} (2d_k)^s \psi^{(s-1)}(d_k m + \eta_k) \right\}.$$

Let Z be the standardized statistic defined by

$$Z = \frac{\widetilde{T} - \kappa^{(1)}}{(\kappa^{(2)})^{1/2}}, \tag{6.12}$$

and let $\widetilde{\kappa}^{(s)}$ be the standardized cumulants defined by $\widetilde{\kappa}^{(s)} = \kappa^{(s)}/(\kappa^{(s)})^{s/2}$. The characteristic function of Z can be expanded as

$$\begin{aligned} C_Z(t) = \mathrm{E}[\exp(itZ)] &= \exp\left\{ -\frac{t^2}{2} + \sum_{s=3}^{\infty} \frac{\widetilde{\kappa}^{(s)}}{s!}(it)^s \right\} \\ &= \exp\left(-\frac{t^2}{2}\right) \left\{ 1 + \sum_{k=1}^{\infty} \frac{1}{k!}(it)^{3k} \left(\sum_{s=0}^{\infty} \frac{\widetilde{\kappa}^{(s+3)}}{(s+3)!}(it)^s \right)^k \right\} \\ &= \exp\left(-\frac{t^2}{2}\right) \left\{ 1 + \sum_{k=1}^{\infty} \frac{1}{k!}(it)^{3k} \sum_{j=0}^{\infty} \gamma_{k,j}(it)^j \right\}, \end{aligned} \tag{6.13}$$

where

$$\gamma_{k,j} = \sum_{s_1+\cdots+s_k=j} \frac{\widetilde{\kappa}^{(s_1+3)} \ldots \widetilde{\kappa}^{(s_k+3)}}{(s_1+3)! \cdots (s_k+3)!}. \tag{6.14}$$

If $\gamma_{k,j} = \mathrm{O}(m^{-(j+k)})$, we can get an approximation for $C_Z(t)$. Therefore, let

$$C_{Z,s}(t) = \exp\left(-\frac{t^2}{2}\right) \left\{ 1 + \sum_{k=1}^{s} \frac{1}{k!}(it)^{3k} \sum_{j=0}^{s-k} \gamma_{k,j}(it)^j \right\}, \tag{6.15}$$

then

$$C_Z(t) = C_{Z,s}(t) + \mathrm{O}(m^{-(s+1)}), \qquad m \to \infty.$$

Inverting the characteristic function formally, we obtain the Edgeworth expansion of the distribution function of Z up to order $\mathrm{O}(m^{-s})$ as

$$Q_s(x) = \Phi(x) - \phi(x) \left\{ \sum_{k=1}^{s} \frac{1}{k!} \sum_{j=0}^{s-k} \gamma_{k,j} H_{3k+j-1}(x) \right\}, \tag{6.16}$$

where Φ and φ are the distribution and density functions of the standard normal distribution, respectively; $\gamma_{k,j}$ is given by (6.15); and $H_r(x)$ is the rth-order Hermite polynomial defined as in Sect. 2.3.1 and (4.20).

As the special cases of (6.16), we give the first-, second-, and third-order asymptotic expansions:

$$\begin{aligned}
Q_0(x) &= \Phi(x), \\
Q_1(x) &= \Phi(x) - \frac{1}{6}\phi(x)\tilde{\kappa}^{(3)}(x^2-1), \\
Q_2(x) &= \Phi(x) - \Phi(x)\Big\{\frac{1}{6}\tilde{\kappa}^{(3)}(x^2-1) + \frac{1}{24}\tilde{\kappa}^{(4)}(x^3-3x) \\
&\quad + \frac{1}{72}(\tilde{\kappa}^{(3)})^2(x^5-10x^3+15x)\Big\}.
\end{aligned} \tag{6.17}$$

Note that the cumulants $\tilde{\kappa}^{(s)}$ involve the digamma and polygamma functions. However, these functions may be computed by, for example, Mathematica Version 5.2.

6.4 Error Bound

In this section we consider the standardized statistic Z defined by (6.12). The characteristic function $C_Z(t)$ of Z can be approximated as $C_{Z,s}(t)$ in (6.16). The inversion of $C_{Z,s}(t)$ gives an asymptotic approximation $Q_s(x)$ in (6.16). Using the inverse Fourier transformation, it is expected to get a uniform bound for the error of the asymptotic expansion above as

$$\begin{aligned}
\sup_x |\Pr\{Z \le x\} - Q_s(x)| &\le \frac{1}{2\pi}\int_{-\infty}^{\infty} \frac{1}{|t|}\left|C_Z(t) - C_{Z,s}(t)\right| dt \\
&= \frac{1}{2\pi}(I_1[v] + I_2[v] + I_3[v]),
\end{aligned} \tag{6.18}$$

where

$$\begin{aligned}
I_1[v] &= \int_{-mv}^{mv} \frac{1}{|t|}\left|C_Z(t) - C_{Z,s}(t)\right| dt, \\
I_2[v] &= \int_{|t|>mv} \frac{1}{|t|}\left|C_{Z,s}(t)\right| dt, \quad \text{and} \quad I_3[v] = \int_{|t|>mv} \frac{1}{|t|}|C_Z(t)|\, dt
\end{aligned}$$

with some positive constant $v < 1$. To find a bound for each integral I_1, I_2, and I_3, we need to evaluate $I_i[v]$, $i = 1, 2, 3$.

For some special cases, a bound for each integral I_1, I_2, and I_3 has been obtained. These include likelihood ratio test statistics for the following:

(1) MANOVA test [9],
(2) independence test [1],
(3) intraclass correlation structure test [3],
(4) additional information test [11],
(5) sphericity test [10],
(6) specified covariance matrix test [10], etc.

In the following we give an outline for deriving error bounds for $I_i[v], i = 1, 2, 3$ in the case of $\widetilde{T} = -\log \Lambda$, based on [9]. In this case,

$$\begin{aligned}
\log \mathrm{E}[\exp(it\widetilde{T})] &= \log \mathrm{E}[\Lambda^{-it}] \\
&= \sum_{j=1}^{q} \left\{ \Gamma\left[\frac{1}{2}(n-p+j) - it\right] - \log \Gamma\left[\frac{1}{2}(n-p+j)\right] \right. \\
&\quad \left. - \left(\log \Gamma\left[\frac{1}{2}(n+j)\right] - \log \Gamma\left[\frac{1}{2}(n+j)\right]\right)\right\} \\
&= \sum_{s=1}^{\infty} \frac{\kappa^{(s)}}{s!}(it)^s,
\end{aligned}$$

where

$$\kappa^{(s)} = (-1)^s \sum_{j=1}^{q} \left\{ \psi^{(s-1)}\left(\frac{1}{2}(n-p+j)\right) - \psi^{(s-1)}\left(\frac{1}{2}(n+j)\right)\right\}. \tag{6.19}$$

Now, consider the standardized statistic $Z = (\widetilde{T} - \kappa^{(1)})/(\kappa^{(2)})^{1/2})$ defined as in (6.12). Here we use the same notation for Z in (6.12), but its cumulants are the ones defined by (6.19). The characteristic function of Z can be expanded as in (6.13) with the cumulants in (6.19). We denote $C_Z(t)$ and $C_{Z,s}(t)$ as $C(t)$ and $C_s(t)$, respectively.

Let m and b_s $(s = 0, 1, 2, \ldots)$ be defined by

$$\begin{aligned}
m &= \frac{n-p-\frac{1}{2}}{2}(\kappa^{(2)})^{1/2}, \\
b_s &= \frac{2}{(s+3)(s+2)(s+1)}(\kappa^{(2)})^{-1} \\
&\quad \times \left\{ 1 - \left(\frac{n-p-\frac{1}{2}}{n-p+q-\frac{1}{2}}\right)^{s+1} - \left(\frac{n-p-\frac{1}{2}}{n-\frac{1}{2}}\right)^{s+1} + \left(\frac{n-p-\frac{1}{2}}{n+q-\frac{1}{2}}\right)^{s+1}\right\}.
\end{aligned}$$

First we derive a bound for $I_1[v]$. Let

$$L[v] = \begin{cases} \dfrac{3v-2}{4v} - \dfrac{(1-v)^2}{2v^2}\log(1-v), & 0 < |v| < 1, \\ 0, & v = 0, \end{cases}$$

$$B[v] = \sum_{s=0}^{\infty} b_s v^s.$$

Then it is easily checked that

$$L[v] = \sum_{s=1}^{\infty} \frac{1}{s(s+1)(s+2)} v^s,$$

$$B[v] = \frac{2}{v\kappa^{(2)}} \left\{ L[v] - L\left[\frac{n-p-\frac{1}{2}}{n-p+q-\frac{1}{2}} v\right] - L\left[\frac{n-p-\frac{1}{2}}{n-\frac{1}{2}} v\right] + L\left[\frac{n-p-\frac{1}{2}}{n+q-\frac{1}{2}} v\right]\right\}.$$

Let

$$R_{k,l}[v] = v^{-l}\left\{ (B[v])^k - \sum_{j=0}^{l-1}\left(\sum_{s_1+\cdots+s_k=j} b_{s_1}\ldots b_{s_k}\right) v^j \right\}.$$

Note that the second term in the above braces is the Taylor expansion of $(B[v])^k$ up to the order v^{l-1}. Then, using $0 < \widetilde{\kappa}^{(s)} < s!\, m^{-(s-2)} b_{s-3}$ $(s = 3, 4, \ldots)$ and (6.13), it is shown that if $|t| \le mv$,

$$\frac{1}{|t|}|C(t) - C_s(t)| \le m^{-(s+1)} \exp\left(-\frac{t^2}{2}\right) \times \left\{ \sum_{k=1}^{s} \frac{1}{k!}|t|^{s+2k} R_{k,s-k+1}[v] + \frac{1}{(s+1)!}|t|^{3s+2}(B[v])^{s+1} \exp\left(t^2 v B[v]\right)\right\}.$$

Integrating the above expression, we obtain a bound for $I_1[v]$:

$$U_1[v] = \frac{2}{m^{s+1}} \left\{ \sum_{k=1}^{s} \frac{1}{k!} R_{k,s-k+1}[v] \int_0^{mv} t^{s+2k} \exp\left(-\frac{t^2}{2}\right) dt + \frac{1}{(s+1)!}(B[v])^{s+1} \int_0^{mv} t^{3s+2} \exp\left(-\frac{t^2}{2} c_v\right) dt \right\},$$

where $c_v = 1 - 2vB[v]$. Note that $U_1[v] = O(m^{-(s+1)})$ if $c_v > 0$.

Next we consider $I_2[v]$ which is represented as

$$I_2[v] = 2\left\{\int_{mv}^{\infty} \exp\left(-\frac{t^2}{2}\right) t^{-1} dt + \sum_{k=1}^{s} \frac{1}{k!} \sum_{j=0}^{s-k} \gamma_{k,j} \int_{mv}^{\infty} \exp\left(-\frac{t^2}{2}\right) |t|^{3k+j-1} dt\right\}.$$

Using

$$2\int_{mv}^{\infty} t^k \exp\left(-\frac{t^2}{2}\right) dt = 2\int_{mv}^{\infty} t^k \exp\left(-\frac{t^2}{2}(1-c+c)\right) dt$$
$$< \exp\left(-\frac{m^2v^2}{2}(1-c)\right)\left(\frac{c}{2}\right)^{-(k+1)/2} \Gamma\left[\frac{k+1}{2}\right],$$

where $0 < c < 1$, we have

$$I_2[v] < \exp\left(-\frac{m^2v^2}{2}(1-c)\right) \times \left\{1 + \sum_{k=1}^{s} \frac{1}{k!} \sum_{j=0}^{s-k} \gamma_{k,j} \left(\frac{c}{2}\right)^{-(3k+j)/2} \Gamma\left[\frac{3k+j}{2}\right]\right\},$$

whose order is $O(\exp[-\frac{m^2v^2}{2}(1-c)])$ $(m \to \infty)$ for fixed v $(0 < v)$ and c $(0 < c < 1)$.

Finally we derive a bound for $I_3[v]$.

The characteristic function of $-\log \Lambda/(\kappa^{(2)})^{1/2}$ is given by

$$\varphi(t) = \prod_{j=1}^{q} \frac{\Gamma[(n-p+j)/2 - i\tilde{t}\,]\Gamma[(n+j)/2]}{\Gamma[(n-p+j)/2]\Gamma[(n+j)/2 - i\tilde{t}\,]},$$

where $\tilde{t} = (\kappa^{(2)})^{-1/2}t$. We use the following properties:

(i) For any real number x, y, $(x > 0)$,

$$\left|\frac{\Gamma[x+yi]}{\Gamma[x]}\right|^2 = \prod_{k=0}^{\infty}\left\{1 + \frac{y^2}{(x+k)^2}\right\}^{-1} \qquad (6.20)$$

(ii) If $a < b$,

$$\log\left\{1 + t^2/(a+x)^2\right\} - \log\left\{1 + t^2/(b+x)^2\right\}$$

is a decreasing function of $x > 0$. Using these properties, the following result is obtained.

$$\log|\varphi(t)| = -\frac{1}{2}\sum_{j=1}^{q}\sum_{k=0}^{\infty}\left\{\log\left(1+\frac{\tilde{t}^2}{(\frac{n-p+j}{2}+k)^2}\right)-\log\left(1+\frac{\tilde{t}^2}{(\frac{n+j}{2}+k)^2}\right)\right\}$$
$$< -\frac{1}{4}\int_1^{q+1}\left\{\int_{n-p}^{n}\log\left(1+\frac{4\tilde{t}^2}{(x+y)^2}\right)dx\right\}dy = -\tilde{t}^2\, G(\tilde{t}, n, p, q),$$

where

$$G(t,n,p,q) = F\left(\frac{n+q+1}{2t}\right) - F\left(\frac{n+1}{2t}\right) - F\left(\frac{n-p+q+1}{2t}\right) + F\left(\frac{n-p+1}{2t}\right),$$
$$F(x) = \int_0^x\left\{\int_0^y \log\left(1+\frac{1}{z^2}\right)dz\right\}dy = \frac{x^2}{2}\log\left(1+\frac{1}{x^2}\right) + 2x\arctan x - \frac{1}{2}\log(1+x^2).$$

Therefore, we obtain a bound for $I_3[v]$ given by

$$I_3[v] < \int_{m_0 v}^{\infty}\frac{2}{t}\exp\left\{-t^2\, G(t,n,p,q)\right\}dt = U_3[v], \tag{6.21}$$

where $m_0 = \frac{1}{2}(n-p-\frac{1}{2})$.

References

1. Akita, T., Jin, J., & Wakaki, H. (2010). High-dimensional Edgeworth expansion of a test statistics on independence and its error bound. *Journal of Multivariate Analysis*, *101*, 1806–1813.
2. Anderson, T. W. (2003). *An introduction to multivariate statistical analysis* (3rd ed.). Hoboken: Wiley.
3. Kato, N., Yamada, T., & Fujikoshi, Y. (2010). High-dimensional asymptotic expansion of LR statistic for testing the intraclass correlation structure and its error bound. *Journal of Multivariate Analysis*, *101*, 101–112.
4. Fujikoshi, Y., & Ulyanov, V. V. (2006). On accuracy of asymptotic expansions for Wilks' lambda distribution. *Journal of Multivariate Analysis*, *97*, 1941–1957.
5. Fujikoshi, Y., Ulyanov, V. V., & Shimizu, R. (2010). *Multivariate statistics: High-dimensional and large-sample approximations*. Hobeken: Wiley.
6. Muirhead, R. J. (1982). *Aspect of multivariate statistical theory*. New York: Wiley.
7. Siotani, M., Hayakawa, T., & Fujikoshi, Y. (1985). *Modern multivariate statistical analysis: A graduate course and handbook*. Columbus: American Sciences Press.

8. Tonda, T., & Fujikoshi, Y. (2004). Asymptotic expansion of the null distribution of LR statistic for multivariate linear hypothesis when the dimension is large. *Communications in Statistics-Theory and Methods A*, *33*, 1205–1220.
9. Wakaki, H. (2007). An error bound for high-dimensional Edgeworth expansion for Wilks' Lambda distribution. *Hiroshima Statistical Research Group*, TR; 7–3.
10. Wakaki, H. (2008, June 19–23). Error bounds for high-dimensional Edgeworth expansions for some tests on covariance matrix. In *International Conference on Multivariate Statistical Modeling and High Dimensional Data Mining*, Kayeseri, Turkey.
11. Wakaki, H., & Fujikoshi, Y. (2017). Computable error bounds for high-dimensional approximations of LR test for additional information in canonical correlation analysis. *Theory of Probability and Its Applications*, *62*, 194–211.

Chapter 7
Bootstrap Confidence Sets

Abstract A sample $X_1, \ldots, X_n$ consisting of independent identically distributed random vectors in $\mathbb{R}^p$ with zero mean and covariance matrix Σ is considered. The recovery of spectral projectors of high-dimensional covariance matrices from a sample of observations is a key problem in statistics arising in numerous applications. This chapter describes a bootstrap procedure for constructing confidence sets for the spectral projector $\mathbf{P}_r$ related to rth eigenvalue of the covariance matrix Σ from given data on the base of corresponding spectral projector $\widehat{\mathbf{P}}_r$ of the sample covariance matrix $\widehat{\Sigma}$. This approach does not use the asymptotical distribution of $\|\mathbf{P}_r - \widehat{\mathbf{P}}_r\|_2$ and does not require the computation of its moment characteristics. The performance of the bootstrap approximation procedure is analyzed.

7.1 Introduction

Consider a sequence of independent identically distributed (i.i.d.) random vectors $X, X_1, \ldots, X_n$, taking values in $\mathbb{R}^p$. Assume that $\mathrm{E}(X) = 0$ and $\mathrm{E}(\|X\|^2) < \infty$. Let $\Sigma \stackrel{\text{def}}{=} \mathrm{E}(XX^\top)$ be the covariance matrix of the vector X. Along with the true covariance matrix Σ we consider the sample covariance matrix $\widehat{\Sigma}$, constructed using the observations $X_1, \ldots, X_n$:

$$\widehat{\Sigma} \stackrel{\text{def}}{=} \frac{1}{n} \sum_{j=1}^{n} X_j X_j^\top = \frac{1}{n} \mathbf{X}\mathbf{X}^\top.$$

Here and below, $\mathbf{X} \stackrel{\text{def}}{=} [X_1, \ldots, X_n] \in \mathbb{R}^{p \times n}$.

In most statistical applications, the true covariance matrix Σ is typically unknown and replaced by its sample counterpart $\widehat{\Sigma}$. The accuracy of estimating Σ by $\widehat{\Sigma}$, in particular, when p is much greater than n, has been extensively studied (see [9–11]). A bound in terms of *the effective rank*,

$$\mathrm{effr}(\Sigma) \stackrel{\text{def}}{=} \mathrm{tr}\,\Sigma / \|\Sigma\|$$

Y. Fujikoshi and V. V. Ulyanov, *Non-Asymptotic Analysis of Approximations for Multivariate Statistics*, JSS Research Series in Statistics,
https://doi.org/10.1007/978-981-13-2616-5_7

was obtained recently in [5]. Such bounds can be used to recover individual eigenvalues of Σ in the case when there are spectral gaps between these eigenvalues.

In this chapter, we recover spectral projectors onto the subspace spanned by eigenvectors corresponding to certain eigenvalues of Σ. The recovery of spectral projectors, as well as eigenvectors and eigenspaces of high-dimensional covariance matrices from a sample of observations is a key problem in statistics that is directly related to dimensionality reduction. For example, in principal component analysis, high-dimensional data are projected onto the subspaces spanned by eigenvectors corresponding to the largest eigenvalues. However, the problem of recovering spectral projectors of high-dimensional covariance matrices has been insufficiently investigated. In [6] non-asymptotic bounds for the Frobenius norm $\|\mathbf{P}_r - \widehat{\mathbf{P}}_r\|_2$ of the distance between sample and true projectors were obtained and asymptotic behavior in large sample size contexts was studied. According to [6], given the moments of observed random variables, one can construct asymptotic confidence sets for the true projector $\mathbf{P}_r$. However, these moment characteristics are typically unknown. On the other hand, it is well known that such asymptotic results can be applied only where the sample size is very large. This is due, in particular, to the fact that the normalized U–statistics arising in the given problem converge extremely slowly to a limiting law.

First we introduce the concept of a spectral gap and define the sample counterpart of the spectral projector $\mathbf{P}_r$. Let σ_j, $j = 1, \dots, p$ denote the eigenvalues of Σ arranged in nonincreasing order and $\mathbf{u}_j$, $j = 1, \dots, p$, be the corresponding eigenvectors. The matrix Σ can be written in terms of its spectral decomposition, namely,

$$\Sigma = \sum_{j=1}^{p} \sigma_j \mathbf{u}_j \mathbf{u}_j^{\mathsf{T}}.$$

Now, let μ_j, $j = 1, \dots, q$, $q \le p$ be different eigenvalues of Σ and $\mathbf{P}_r$, $r = 1, \dots, q$ be the corresponding spectral projectors, i.e.,

$$\mathbf{P}_r = \sum_{j:\, \sigma_j = \mu_r} \mathbf{u}_j \mathbf{u}_j^{\mathsf{T}}.$$

Then

$$\Sigma = \sum_{r=1}^{q} \mu_r \mathbf{P}_r.$$

Let $\Delta_r \stackrel{\text{def}}{=} \{j \colon \sigma_j = \mu_r\}$. It is easy to see that $|\Delta_r| = m_r$, where $m_r \stackrel{\text{def}}{=} \operatorname{rank}(\mathbf{P}_r)$. For all $r \ge 1$, we define $g_r \stackrel{\text{def}}{=} \mu_r - \mu_{r+1} > 0$. Let $\overline{g}_r \stackrel{\text{def}}{=} \min(g_{r-1}, g_r)$ for $r \ge 2$ and $\overline{g}_1 \stackrel{\text{def}}{=} g_1$. Here, $\overline{g}_r$ is called the rth spectral gap corresponding to the eigenvalue μ_r.

Now we write $\widehat{\Sigma}$ in terms of its spectral decomposition:

$$\widehat{\Sigma} = \sum_{j=1}^{p} \widehat{\sigma}_j \widehat{\mathbf{u}}_j \widehat{\mathbf{u}}_j^{\mathsf{T}},$$

where $\widehat{\sigma}_1 \geq \widehat{\sigma}_2 \geq \cdots \geq \widehat{\sigma}_p, \widehat{\mathbf{u}}_1, \ldots, \widehat{\mathbf{u}}_p$ are the eigenvalues and corresponding eigenvectors of the matrix $\widehat{\Sigma}$. Following [6], we define *clusters* of eigenvalues $\widehat{\sigma}_j$, $j \in \Delta_r$. Let $\widehat{\mathbf{E}} \overset{\text{def}}{=} \widehat{\Sigma} - \Sigma$. It is straightforward that

$$\inf_{j \notin \Delta_r} |\widehat{\sigma}_j - \mu_r| \geq \overline{g}_r - \|\widehat{\mathbf{E}}\|, \quad \sup_{j \in \Delta_r} |\widehat{\sigma}_j - \mu_r| \leq \|\widehat{\mathbf{E}}\|.$$

Assume that $\|\widehat{\mathbf{E}}\| \leq \overline{g}_r/2$. Then all $\widehat{\sigma}_j$, $j \in \Delta_r$ belong to the interval

$$(\mu_r - \|\widehat{\mathbf{E}}\|, \mu_r + \|\widehat{\mathbf{E}}\|) \subset \big(\mu_r - \overline{g}_r/2, \mu_r + \overline{g}_r/2\big),$$

while the other eigenvalues of $\widehat{\Sigma}$ lie outside the interval

$$\Big(\mu_r - (\overline{g}_r - \|\widehat{\mathbf{E}}\|), \mu_r + (\overline{g}_r - \|\widehat{\mathbf{E}}\|)\Big) \supset \big[\mu_r - \overline{g}_r/2, \mu_r + \overline{g}_r/2\big].$$

If we additionally require that

$$\|\widehat{\mathbf{E}}\| < \frac{1}{4} \min_{1 \leq s \leq r} \overline{g}_s =: \overline{\delta}_r,$$

then the set $\{\widehat{\sigma}_j, j \in \cup_{s=1}^{r} \Delta_s\}$ consists of r clusters. The diameter of each cluster is strictly less than $2\overline{\delta}_r$, and the distance between any two clusters is greater than $2\overline{\delta}_r$. Let $\widehat{\mathbf{P}}_r$ denote the projector onto the subspace spanned by $\widehat{\mathbf{u}}_j$, $j \in \Delta_r$.

As was noted above, the asymptotic normality of $\|\widehat{\mathbf{P}}_r - \mathbf{P}_r\|_2^2$ was proved in [6], Theorem 6. Relying on this result, one can construct asymptotic confidence sets for an unknown projector $\mathbf{P}_r$ of the form

$$\left\{\mathbf{P}_r : \frac{\|\widehat{\mathbf{P}}_r - \mathbf{P}_r\|_2^2 - \mathrm{E}\|\widehat{\mathbf{P}}_r - \mathbf{P}_r\|_2^2}{\mathrm{Var}^{1/2}(\|\widehat{\mathbf{P}}_r - \mathbf{P}_r\|_2^2)} \leq z_\alpha\right\},$$

where z_α is the α-quantile of the normal distribution. The fundamental drawbacks of this approach are the slow convergence rate to the normal law and the fact that a large sample size is required to achieve a reasonable approximation. Additionally, we need to estimate $\mathrm{E}(\|\widehat{\mathbf{P}}_r - \mathbf{P}_r\|_2^2)$ and $\mathrm{Var}(\|\widehat{\mathbf{P}}_r - \mathbf{P}_r\|_2^2)$, which depend on the unknown matrix Σ. A partial solution of this problem is discussed in [6]. It involves splitting the sample into three subsamples, and pilot estimation of the expectation and variance of $\|\widehat{\mathbf{P}}_r - \mathbf{P}_r\|_2^2$. The approach only applies in some special cases, in particular, if the covariance matrix has a nearly spike structure.

7.2 Bootstrap Procedure

In this section, the quantile

$$\gamma_\alpha \stackrel{\text{def}}{=} \inf\left\{\gamma > 0\colon \Pr\left\{n\|\widehat{\mathbf{P}}_r - \mathbf{P}_r\|_2^2 > \gamma\right\} \le \alpha\right\} \tag{7.1}$$

is estimated using a bootstrap procedure without estimating the covariance matrix Σ. Moreover, this approach has the following properties:

(i) it does not rely on the asymptotic distribution of $\|\widehat{\mathbf{P}}_r - \mathbf{P}_r\|_2^2$;
(ii) it does not require computing the moments of $\|\widehat{\mathbf{P}}_r - \mathbf{P}_r\|_2^2$;
(iii) it does not require splitting the sample into subsamples; and
(iv) it provides an explicit error bound for the bootstrap approximation.

Note that the bootstrap method is one of the most widespread statistical techniques for constructing confidence sets. However, existing theory proves the possibility of applying this method to parametric models. Generalization to the case when the space dimension is much higher than the sample size encounters various difficulties. In this context, we note [2, 3, 8] . Herein, the bootstrap method is extended to the construction of confidence sets for spectral projectors. Additionally, it should be emphasized that spectral projectors depend nonlinearly on the covariance matrix, which in turn is a quadratic function of the original distribution, and this causes further difficulties.

The weighted (or *bootstrap*) version of the matrix $\widehat{\Sigma}$ is defined as

$$\Sigma^\circ \stackrel{\text{def}}{=} \frac{1}{n}\sum_{i=1}^{n} w_i X_i X_i^{\mathsf{T}},$$

where $w_1, \dots, w_n$ are i.i.d. random variables independent of $\mathbf{X} = (X_1, \dots, X_n)$, such that $\mathrm{E}\,(w_1) = 1$, $\mathrm{Var}\,(w_1) = 1$. As an example, we can consider i.i.d. Gaussian weights $w_i \sim \mathrm{N}(1, 1)$. We introduce the conditional probability $\Pr^\circ\{\cdot\} \stackrel{\text{def}}{=} \Pr\{\cdot \mid \mathbf{X}\}$ and denote the corresponding expectation by E°. It is straightforward that, if the sample is fixed and the only random variables are the weights w_i, then the expectation Σ° is the known matrix $\widehat{\Sigma}$, i.e., $\mathrm{E}^\circ(\Sigma^\circ) = \widehat{\Sigma}$. This is opposite to $\mathrm{E}(\widehat{\Sigma}) = \Sigma$, where Σ is unknown.

Writing

$$\Sigma^\circ = \sum_{j=1}^{p} \sigma_j^\circ \mathbf{u}_j^\circ \mathbf{u}_j^{\circ\mathsf{T}},$$

we define the spectral projectors $\mathbf{P}_r^\circ$ as the orthogonal projectors onto the subspace spanned by $\mathbf{u}_j^\circ$, $j \in \Delta_r$.

For a given α, the quantile γ_α° is defined as (cf. (7.1))

$$\gamma_\alpha^\circ \stackrel{\text{def}}{=} \min\left\{\gamma > 0\colon \Pr^\circ\left\{n\|\mathbf{P}_r^\circ - \widehat{\mathbf{P}}_r\|_2^2 > \gamma\right\} \le \alpha\right\}. \tag{7.2}$$

Note that γ_α° depends on the sample $\mathbf{X}$. The essence of the method is to use γ_α° for constructing the confidence set

$$\mathcal{E}(\alpha) \stackrel{\text{def}}{=} \{\mathbf{P}\colon n\|\mathbf{P} - \widehat{\mathbf{P}}_r\|_2^2 \le \gamma_\alpha^\circ\}.$$

Thus, we need to show that

$$\Pr\{\mathbf{P}_r \notin \mathcal{E}(\alpha)\} = \Pr\{n\|\mathbf{P}_r - \widehat{\mathbf{P}}_r\|_2^2 > \gamma_\alpha^\circ\} \approx \alpha.$$

7.3 Confidence Sets for Spectral Projectors: Bootstrap Validity

Define the block matrix

$$\Gamma_r \stackrel{\text{def}}{=} \begin{pmatrix} \Gamma_{r1} & \mathbf{O} & \cdots & \mathbf{O} \\ \mathbf{O} & \Gamma_{r2} & \mathbf{O} \cdots & \mathbf{O} \\ \cdots & & & \\ \mathbf{O} & \cdots & \mathbf{O} & \Gamma_{rq} \end{pmatrix}, \tag{7.3}$$

where Γ_{rs}, $s \ne r$ are diagonal $m_r m_s \times m_r m_s$ matrices with values $2\mu_r\mu_s/(\mu_r - \mu_s)^2$ on the main diagonal. Let $\lambda_1(\Gamma_r) \ge \lambda_2(\Gamma_r) \ge \cdots$ denote the eigenvalues of Γ_r. According to available bounds for the distance between Σ and $\widehat{\Sigma}$, the eigenvalues of Σ can be recovered with accuracy $\mathrm{O}(1/\sqrt{n})$. Thus, the part of the spectrum of Σ below a threshold of order $\mathrm{O}(1/\sqrt{n})$ cannot be estimated. The same is true of the matrix Γ_r. The value m is defined by the following inequalities:

$$\lambda_{\mathrm{m}}(\Gamma_r) \ge \operatorname{tr}\Gamma_r \left(\sqrt{\frac{\log n}{n}} + \sqrt{\frac{\log p}{n}}\right) > \lambda_{\mathrm{m}+1}(\Gamma_r). \tag{7.4}$$

We agree to write below $a \lesssim b$ ($a \gtrsim b$) if there exists some absolute constant C such that $a \le Cb$ ($a \ge Cb$ resp.). Let Π_{m} denote the projector onto the subspace spanned by the eigenvectors of Γ_r corresponding to its largest m eigenvalues. The following theorem is the main result of this chapter.

Theorem 7.1 *Assume that observations $X, X_1, \ldots, X_n$ are i.i.d. Gaussian random vectors in $\mathbb{R}^p$ with zero mean and covariance matrix $\mathrm{E}(XX^\top) = \Sigma$. For any α : $0 < \alpha < 1$, the corresponding quantile γ_α° is defined by (7.2), where the weights are additionally assumed to satisfy $w_i \sim \mathrm{N}(1, 1)$ for all $i = 1, \ldots, n$. Then*

$$\left|\alpha - \Pr\{n\|\widehat{\mathbf{P}}_r - \mathbf{P}_r\|_2^2 > \gamma_\alpha^\circ\}\right| \lesssim \Delta, \tag{7.5}$$

where

$$\Delta \stackrel{\text{def}}{=} \frac{\mathrm{m}\,\mathrm{tr}\Gamma_r}{\sqrt{\lambda_1(\Gamma_r)\lambda_2(\Gamma_r)}}\left(\sqrt{\frac{\log n}{n}} + \sqrt{\frac{\log p}{n}}\right) + \frac{\mathrm{tr}(\mathbf{I} - \Pi_{\mathrm{m}})\Gamma_r}{\sqrt{\lambda_1(\Gamma_r)\lambda_2(\Gamma_r)}}$$
$$+ \frac{m_r(\mathrm{tr}\,\Sigma)^3}{\overline{g}_r^3\sqrt{\lambda_1(\Gamma_r)\lambda_2(\Gamma_r)}}\left(\sqrt{\frac{\log^3 n}{n}} + \sqrt{\frac{\log^3 p}{n}}\right)$$

and m *is defined by (7.4).*

Remark 7.1 To replace the Gaussian assumption for $X_1, \ldots, X_n$ by more realistic setup, for example, sub-Gaussian or -exponential assumption is a challenge for a future research. Among other difficulties it will require, in particularly, a version of the central limit theorem for non i.i.d. elements in high-dimensional space with precise dependence of the rate of convergence on the dimension p. Some partial results are available, e.g., [1], but they provide dependence on p, which is not sufficient for our purposes.

Remark 7.2 We choose w_j, $j = 1, \ldots, n$ to be Gaussian r.v. variables. This choice may lead to the situation that the matrix Σ° may have negative eigenvalues. This is not critical problem since we are mostly interested in the largest eigenvalues. Assuming that the rth gap $\overline{g}_r > 0$ we get that the rth largest eigenvalue of Σ° is positive and concentrated around μ_r with high probability.

Remark 7.3 The result (7.5) implicitly assumes that the error term Δ is small. If $\Delta \geq 1$ then (7.5) is meaningless. In particular, this implies that $p \lesssim \exp(n^{1/3})$.

Remark 7.4 The error term Δ can be described in terms of Σ. It is easy to check that for all r

$$\mathrm{tr}\Gamma_r \lesssim m_r \frac{\mu_r \mathrm{tr}\,\Sigma}{\overline{g}_r^2} \leq m_r \frac{\|\Sigma\|^2 \mathrm{effr}(\Sigma)}{\overline{g}_r^2}.$$

Let us consider, for example, the case $r = 2$ and $m_1 = m_2 = 1$. Introduce a function

$$f(x) = 2x\mu_2/(x - \mu_2)^2$$

at the points $x = \mu_s$, $s \neq 2$. It is straightforward to check that the maximum of $f(x)$ is achieved at $x = \mu_1$ or μ_3. Moreover, assume that the largest values of $f(x)$ are $f(\mu_1)$ and $f(\mu_3)$. Then

$$\Delta \lesssim \frac{\mathrm{m}\,\mathrm{tr}\,\Sigma}{\overline{g}_2}\sqrt{\frac{\mu_1}{\mu_3}}\left(\sqrt{\frac{\log n}{n}} + \sqrt{\frac{\log p}{n}}\right) + \sqrt{\frac{\mu_1}{\mu_3}}\frac{\mathrm{tr}(\mathbf{I} - \Pi_{\mathrm{m}})\Sigma}{\overline{g}_2}$$
$$+ \frac{\mathrm{tr}^3\,\Sigma}{\overline{g}_2^2\,\mu_2}\sqrt{\frac{\mu_1}{\mu_3}}\left(\sqrt{\frac{\log^3 n}{n}} + \sqrt{\frac{\log^3 p}{n}}\right).$$

Table 7.1 Coverage probabilities. For each n the first line corresponds to the median value of the coverage probability and the second line corresponds to the interquartile range

n	Confidence levels					
	0.99	0.95	0.90	0.85	0.80	0.75
100	0.992	0.961	0.918	0.876	0.825	0.768
	0.027	0.091	0.146	0.197	0.231	0.257
300	0.988	0.942	0.886	0.832	0.784	0.735
	0.020	0.062	0.094	0.118	0.139	0.153
500	0.995	0.966	0.925	0.876	0.822	0.771
	0.013	0.035	0.072	0.104	0.120	0.122
1000	0.989	0.957	0.906	0.848	0.795	0.743
	0.012	0.038	0.062	0.086	0.093	0.098

Remark 7.5 The performance of the bootstrap procedure can be illustrated by numerical examples. Namely, we check how well is the bootstrap approximation of the true quantiles.

Let n be a sample size. We consider the different values of n, namely, $n = 100, 300, 500, 1000$. Let $X, X_1, \ldots, X_n$ have the normal distribution in $\mathbb{R}^p$, with zero mean and covariance matrix Σ. We take $p = 100$ and Σ is the diagonal matrix with the eigenvalues $\mu_1 = 25.698$, $\mu_2 = 15.7688$, $\mu_3 = 10.0907$, $\mu_4 = 5.9214$, $\mu_5 = 3.4321$, and $\mu_6, \ldots, \mu_{100}$ are distributed according to Marchenko–Pastur's density with the support on $[0.71, 1.34]$. Then $\overline{g}_1 = 9, 93$ and $\text{effr}(\Sigma) = 6.12$. The distribution of $n\|\widehat{\mathbf{P}}_1 - \mathbf{P}_1\|_2^2$ is evaluated by using 3000 Monte-Carlo samples from the normal distribution with zero mean and covariance Σ. The bootstrap distribution for a given realization X is evaluated by 3000 Monte-Carlo samples of bootstrap weights $\{w_1, \ldots, w_n\}$. Since this distribution is random and depends on X, we finally use the median from 50 realizations of X for each quantile. The coverage probabilities are collected in Table 7.1.

For other numerical examples and PP-plots, see in Sect. 3 in [7]. It shows that the bootstrap procedure mimics well the most of the underlying distribution of $n\|\widehat{\mathbf{P}}_1 - \mathbf{P}_1\|_2^2$.

Remark 7.6 For details of the proof of Theorem 7.1 see [7]. Here we outline its main steps only.

To prove (7.5) it is sufficient to show that with high probability for all $x > 0$

$$\Pr{}^{\circ}\{n\|\mathbf{P}_r^{\circ} - \widehat{\mathbf{P}}_r\|_2^2 > x\} \approx \Pr\{n\|\widehat{\mathbf{P}}_r - \mathbf{P}_r\|_2^2 > x\}.$$

First we get that for all $x > 0$

$$\Pr\{n\|\widehat{\mathbf{P}}_r - \mathbf{P}_r\|_2^2 > x\} \approx \Pr\{\|\boldsymbol{\xi}\|^2 > x\},$$

where $\boldsymbol{\xi}$ is a Gaussian vector with zero mean and covariance matrix Γ_r, defined in (7.3).

Further, we show that the similar relation with high probability holds in so-called bootstrap world, namely,

$$\Pr{}^\circ\{n\|\mathbf{P}_r^\circ - \widehat{\mathbf{P}}_r\|_2^2 > x\} \approx \Pr{}^\circ\{\|\boldsymbol{\xi}^\circ\|^2 > x\},$$

where $\boldsymbol{\xi}^\circ$ is a Gaussian vector with zero mean and covariance matrix $\Gamma^\circ{}_r$, which is different from Γ_r but close to it.

Therefore, we come to the problem to estimate the closeness (on balls) of two centered Gaussian measures with different covariance operators. A similar problem arises in other statistical contexts, see an example in the beginning of next section. Moreover, in next section, a more general problem is solved; namely, upper bounds are obtained for the closeness of two Gaussian measures with different means and covariance operators in the class of balls in a separable Hilbert space (see [4]). The bounds are optimal with respect to the dependence on the spectra of the covariance operators of Gaussian measures. The inequalities cannot be improved in the general case.

References

1. Bentkus, V. (2005). A Lyapunov-type bound in R^d. *Theory of Probability & Its Applications, 49*(2), 311–323.
2. Chernozhukov, V., Chetverikov, D., & Kato, K. (2013). Gaussian approximations and multiplier bootstrap for maxima of sums of high-dimensional random vectors. *The Annals of Statistics, 41*(6), 2786–2819.
3. Chernozhukov, V., Chetverikov, D., & Kato, K. (2017). Central limit theorems and bootstrap in high dimensions. *The Annals of Probability, 45*(4), 2309–2352.
4. Götze, G., Naumov, A., Spokoiny, V., & Ulyanov, V. (2019). Large ball probabilities, Gaussian comparison and anti-concentration. *Bernoulli, 25*(4A), 2538–2563.
5. Koltchinskii, V., & Lounici, K. (2017). Concentration inequalities and moment bounds for sample covariance operators. *Bernoulli, 23*(1), 110–133.
6. Koltchinskii, V., & Lounici, K. (2017). Normal approximation and concentration of spectral projectors of sample covariance. *The Annals of Statistics, 45*(1), 121–157.
7. Naumov, A., Spokoiny, V., & Ulyanov, V. (2019). Bootstrap confidence sets for spectral projectors of sample covariance. *Probability Theory and Related Fields, 174*(3–4), 1091–1132.
8. Spokoiny, V., & Zhilova, M. (2015). Bootstrap confidence sets under model misspecification. *The Annals of Statistics, 43*(6), 2653–2675.
9. Tropp, J. (2012). User-friendly tail bounds for sums of random matrices. *Foundations of Computational Mathematics, 12*(4), 389–434.
10. van Handel, R. (2017). On the spectral norm of Gaussian random matrices. *Convexity and concentration* (Vol. 161, pp. 107–165)., IMA Berlin: Springer.
11. Vershynin, R. (2012). Introduction to the non-asymptotic analysis of random matrices. *Compressed sensing* (pp. 210–268). Cambridge: Cambridge University Press.

Chapter 8
Gaussian Comparison and Anti-concentration

Abstract We derive tight non-asymptotic bounds for the Kolmogorov distance between the probabilities of two Gaussian elements to hit a ball in a Hilbert space. The key property of these bounds is that they are dimension-free and depend on the nuclear (Schatten-one) norm of the difference between the covariance operators of the elements and on the norm of the mean shift. The obtained bounds significantly improve the bound based on the Pinsker inequality via the Kullback–Leibler divergence. We also establish an anti-concentration bound for a squared norm of a non-centered Gaussian element in a Hilbert space. A number of examples are also provided, motivating the results and its applications to statistical inference and high-dimensional CLT.

8.1 Introduction

In many statistical and probabilistic applications, one faces the problem of Gaussian comparison, i.e., one has to evaluate how the probability of a ball under a Gaussian measure is affected if the mean and covariance operators of this Gaussian measure are slightly changed. In Chap. 7 we gave an example on bootstrap validation when such "large ball probability" problems naturally arise. Other examples, including bootstrap validation, Bayesian inference and high-dimensional CLT, are discussed below. Sharp bounds are presented for the Kolmogorov distance between the probabilities of two Gaussian elements to hit a ball in a Hilbert space. The key property of these bounds is that they are dimension-free and depend on the nuclear (Schatten-one) norm of the difference between the covariance operators of the elements. We also state a tight dimension-free anti-concentration bound for a squared norm of a Gaussian element in Hilbert space which refines the well-known results on the density of a chi-squared distribution.

Through the examples considered, we explain how the bounds of this chapter can be used to improve on existing results. The key observation behind the improvement is that we only need to know the properties of Gaussian measures on a class of balls.

Y. Fujikoshi and V. V. Ulyanov, *Non-Asymptotic Analysis of Approximations for Multivariate Statistics*, JSS Research Series in Statistics,
https://doi.org/10.1007/978-981-13-2616-5_8

This means, in particular, that we would like to compare two Gaussian measures on the class of balls instead of on the class of all measurable sets. The latter can be upper-bounded by a general Pinsker inequality via the Kullback–Leibler divergence. For Gaussian measures, this divergence can be expressed in terms of parameters of the underlying measures; see, e.g., [11]. However, the obtained bound involves the inverse of the covariance operators of the considered Gaussian measures. In particular, small eigenvalues have the largest impact, which is counter-intuitive if a probability of a ball is considered. Our bounds only involve the operator and Frobenius norms of the related covariance operators and apply even in a Hilbert space.

8.2 Motivation: Prior Impact in Linear Gaussian Modeling

We start by considering prior impact in linear Gaussian modeling.

Consider a linear regression model $Y_i = \boldsymbol{X}_i^\top \boldsymbol{\theta} + \varepsilon_i$. The assumption of homogeneous Gaussian errors $\varepsilon_i \sim \mathrm{N}(0, \sigma^2)$ yields the log-likelihood

$$L(\boldsymbol{\theta}) = -\frac{1}{2\sigma^2}\sum_{i=1}^{n}(Y_i - \boldsymbol{X}_i^\top \boldsymbol{\theta})^2 + R = -\frac{1}{2\sigma^2}\bigl\|\mathbf{Y} - \boldsymbol{X}^\top \boldsymbol{\theta}\bigr\|^2 + R,$$

where $\boldsymbol{X} = (\boldsymbol{X}_1, \ldots. \boldsymbol{X}_n)$ and the term R does not depend on $\boldsymbol{\theta}$. A Gaussian prior $\Pi = \Pi_G = \mathrm{N}\bigl(0, G^{-2}\bigr)$ implies that the posterior distribution for the target parameter $\boldsymbol{\theta}$ given $\mathbf{Y}$ can be written as

$$\boldsymbol{\vartheta}_G \,\big|\, \mathbf{Y} \propto \exp\Bigl(L(\boldsymbol{\theta}) - \frac{1}{2}\|G\boldsymbol{\theta}\|^2\Bigr) \propto \exp\Bigl(-\frac{1}{2\sigma^2}\bigl\|\mathbf{Y} - \boldsymbol{X}^\top \boldsymbol{\theta}\bigr\|^2 - \frac{1}{2}\|G\boldsymbol{\theta}\|^2\Bigr).$$

We represent the quantity $L_G(\boldsymbol{\theta}) \stackrel{\text{def}}{=} L(\boldsymbol{\theta}) - \frac{1}{2}\|G\boldsymbol{\theta}\|^2$ in the form

$$L_G(\boldsymbol{\theta}) = L_G(\widehat{\boldsymbol{\theta}}_G) - \frac{1}{2}\bigl\|D_G(\boldsymbol{\theta} - \widehat{\boldsymbol{\theta}}_G)\bigr\|^2,$$

where $\widehat{\boldsymbol{\theta}}_G \stackrel{\text{def}}{=} \bigl(\boldsymbol{X}\boldsymbol{X}^\top + \sigma^2 G^2\bigr)^{-1}\boldsymbol{X}\mathbf{Y}$, $D_G^2 \stackrel{\text{def}}{=} \sigma^{-2}\boldsymbol{X}\boldsymbol{X}^\top + G^2$. In particular, this implies that the posterior distribution $\Pr\{\boldsymbol{\vartheta}_G \,\big|\, \mathbf{Y})\}$ o f $\boldsymbol{\vartheta}_G$ given $\mathbf{Y}$ is $\mathrm{N}(\widehat{\boldsymbol{\theta}}_G, D_G^{-2})$. A contraction property is a kind of concentration of the posterior on the elliptic set:

$$E_G(\mathtt{r}) = \bigl\{\boldsymbol{\theta} \colon \|W(\boldsymbol{\theta} - \widehat{\boldsymbol{\theta}}_G)\| \le \mathtt{r}\bigr\},$$

where W is a given linear mapping from $\mathbb{R}^p$. The desirable credibility property manifests the prescribed conditional probability of $\boldsymbol{\vartheta}_G \in E(\mathtt{r}_G)$ given $\mathbf{Y}$ with $\mathtt{r}_G$ defined for a given α by

$$\Pr\Bigl\{ \bigl\| W\bigl(\vartheta_G - \widehat{\theta}_G\bigr) \bigr\| \ge \mathtt{r}_G \bigm| \mathbf{Y} \Bigr\} = \alpha. \tag{8.1}$$

Under the posterior measure $\vartheta_G \sim \mathrm{N}(\widehat{\theta}_G, D_G^{-2})$, this bound can be expressed as

$$\Pr\bigl\{ \|\xi_G\| \ge \mathtt{r}_G \bigr\} = \alpha \tag{8.2}$$

with a zero mean normal vector $\xi_G \sim \mathrm{N}(\mathbf{0}, \Sigma_G)$ for $\Sigma_G = W D_G^{-2} W^\top$. The issue of prior impact can be stated as follows: whether the obtained credible set significantly depends on the prior covariance G. Consider another prior $\Pi_1 = \mathrm{N}(\mathbf{0}, G_1^{-2})$ with the covariance matrix G_1^{-2}. The corresponding posterior ϑ_{G_1} is again normal, but now with parameters $\widehat{\theta}_{G_1} = \bigl(XX^\top + \sigma^2 G_1^2\bigr)^{-1} X\mathbf{Y}$ and $D_{G_1}^2 = \sigma^{-2} XX^\top + G_1^2$. We seek to check the posterior probability of the credible set $E_G(\mathtt{r}_G)$:

$$\Pr\Bigl\{ \bigl\| W\bigl(\vartheta_{G_1} - \widehat{\theta}_G\bigr) \bigr\| \ge \mathtt{r}_G \bigm| \mathbf{Y} \Bigr\}.$$

Clearly this probability can be written as $\Pr\Bigl\{ \|\xi_{G_1} + \boldsymbol{a}\| \ge \mathtt{r}_G \Bigr\}$ with $\xi_{G_1} \sim \mathrm{N}(0, \Sigma_{G_1})$ for $\Sigma_{G_1} = W D_{G_1}^{-2} W^\top$ and $\boldsymbol{a} \stackrel{\mathrm{def}}{=} W\bigl(\widehat{\theta}_{G_1} - \widehat{\theta}_G\bigr)$. Therefore,

$$\begin{aligned} &\Bigl| \Pr\Bigl\{ \bigl\| W\bigl(\vartheta_{G_1} - \widehat{\theta}_G\bigr) \bigr\| \ge \mathtt{r}_G \bigm| \mathbf{Y} \Bigr\} - \alpha \Bigr| \\ &\qquad \le \sup_{\mathtt{r}>0} \Bigl| \mathrm{P}\Bigl(\|\xi_{G_1} + \boldsymbol{a}\| \ge \mathtt{r} \Bigr) - \Pr\Bigl\{ \|\xi_G\| \ge \mathtt{r} \Bigr\} \Bigr| . \end{aligned}$$

The Pinsker inequality (see, e.g., Sect. 2.4 in [12]) implies an upper bound for the total variation distance, i.e., $\|\cdot\|_{TV}$, between the Gaussian measures $\mathrm{N}(\mathbf{0}, \Sigma_G)$ and $\mathrm{N}(\boldsymbol{a}, \Sigma_{G_1})$; however, the answer is given via the Kullback–Leibler divergence for these two measures in terms of the Frobenius norm, i.e., $\|\cdot\|_{Fr}$:

$$\bigl\| \mathrm{N}(\mathbf{0}, \Sigma_G) - \mathrm{N}(\boldsymbol{a}, \Sigma_{G_1}) \bigr\|_{TV} \le \mathtt{C}\Bigl(\bigl\| \Sigma_G^{-1/2} \Sigma_{G_1} \Sigma_G^{-1/2} - \mathbf{I}_p \bigr\|_{Fr} + \bigl\| \Sigma_{G_1}^{-1/2} \boldsymbol{a} \bigr\| \Bigr); \tag{8.3}$$

see, e.g., [10]. Our results allow us to significantly improve this bound when we need an upper bound not for total variation of two Gaussian measures but for its closeness on the centered balls. In particular, only the nuclear norm $\bigl\| \Sigma_G - \Sigma_{G_1} \bigr\|_1$, the norm of the vector $\boldsymbol{a}$ and the Frobenius norm of Σ_G are involved. If $G^2 \ge G_1^2$, then $\Sigma_G \le \Sigma_{G_1}$ and $\bigl\| \Sigma_G - \Sigma_{G_1} \bigr\|_1 = \operatorname{tr} \Sigma_{G_1} - \operatorname{tr} \Sigma_G$. Thus, by Theorem 8.2 below, it holds under some technical conditions and with an absolute constant C

$$\Bigl| \Pr\Bigl\{ \bigl\| W\bigl(\vartheta_{G_1} - \widehat{\theta}_G\bigr) \bigr\| \ge \mathtt{r}_G \bigm| \mathbf{Y} \Bigr\} - \alpha \Bigr| \le \frac{C\,\bigl(\operatorname{tr} \Sigma_{G_1} - \operatorname{tr} \Sigma_G + \|\boldsymbol{a}\|^2\bigr)}{\|\Sigma_G\|_{Fr}} .$$

This new bound significantly outperforms (8.3).

Other motivating example on bootstrap validity see in the previous Sect. 6. See also Sect. "Application examples" in [6], where it is shown that similar problem about Gaussian comparison arises when considering bootstrap validity for the MLE, nonparametric Bayes approach or Central Limit Theorem in finite- and infinite-dimensional spaces.

8.3 Gaussian Comparison

Let $\boldsymbol{\xi}$ and $\boldsymbol{\eta}$ be two Gaussian vectors in $\mathbb{R}^p$ with covariance matrices $\boldsymbol{\Sigma}_\xi$ and $\boldsymbol{\Sigma}_\eta$, respectively. First for simplicity we assume that $\boldsymbol{\xi}$ and $\boldsymbol{\eta}$ have zero means. We want to estimate

$$\delta(\mathcal{C}) := \sup_{C \in \mathcal{C}} |\Pr\{\boldsymbol{\xi} \in C\} - \Pr\{\boldsymbol{\eta} \in C\}|, \tag{8.4}$$

for some class $\mathcal{C}$ of measurable subsets of $\mathbb{R}^p$. For this purpose, we can use the Pinsker inequality (see, e.g., Sect. 2.4 in [12]), which, in the case of the class of all Borel sets in $\mathbb{R}^p$ taken as $\mathcal{C}$, estimates the total variation distance of measures in terms of the Kullback–Leibler divergence. In the Gaussian case, it can be written explicitly. However, (8.4) frequently needs to be estimated much more accurately, but for a smaller class $\mathcal{C}$. In [1–3], for example, the class of all rectangles in $\mathbb{R}^p$ was considered and optimal estimates were found. Additionally, in statistical applications (bootstrap method or Bayesian analysis) it is very important (see, e.g., [9, 11]) to obtain p-independent estimates of (8.4) for the class of all centered balls $\mathcal{C}_0 := (\{\boldsymbol{x} \in \mathbb{R}^p : \|\boldsymbol{x}\| \le r\}, r \ge 0)$ in $\mathbb{R}^p$. Here and below, $\|\boldsymbol{x}\|$ is the Euclidean vector norm.

We present an estimate for $\delta(\mathcal{C}_0)$ in terms of corresponding covariance operators so that the resulting estimates cannot be improved in the general case. We consider Gaussian random elements in a separable Hilbert space $\mathbb{H}$ with zero means. For a self-adjoint nonnegative linear operator $\mathbf{A}$ in $\mathbb{H}$ with nonincreasing eigenvalues $\lambda_1(\mathbf{A}) \ge \lambda_2(\mathbf{A}), \ldots$, let $\lambda(\mathbf{A})$ denote the diagonal operator $\operatorname{diag}(\lambda_1(\mathbf{A}), \lambda_2(\mathbf{A}), \ldots)$. For a self-adjoint linear operator $\mathbf{B}$ in $\mathbb{H}$, the trace norm is defined as $\|\mathbf{B}\|_1 := \operatorname{tr}|\mathbf{B}| := \sum_{k=1}^{\infty} |\lambda_k(\mathbf{B})|$, where $\lambda_k(\mathbf{B})$ are the eigenvalues of $\mathbf{B}$. We agree to write below $a \lesssim b$ ($a \gtrsim b$) if there exists some absolute constant C such that $a \le Cb$ ($a \ge Cb$ resp.).

The following theorem is the main result of this section.

Theorem 8.1 *Let $\boldsymbol{\xi}$ and $\boldsymbol{\eta}$ be Gaussian elements in $\mathbb{H}$, with zero mean and covariance operators $\boldsymbol{\Sigma}_\xi$ and $\boldsymbol{\Sigma}_\eta$, respectively. Let $\lambda_{1\xi} \ge \lambda_{2\xi} \ge \cdots$ and $\lambda_{1\eta} \ge \lambda_{2\eta} \ge \cdots$ be the respective eigenvalues of $\boldsymbol{\Sigma}_\xi$ and $\boldsymbol{\Sigma}_\eta$. Then*

$$\begin{aligned}\delta(\mathcal{C}_0) &= \sup_{x>0} |\Pr\{\|\boldsymbol{\xi}\| \le x\} - \Pr\{\|\boldsymbol{\eta}\| \le x\}| \\ &\lesssim \left((\Lambda_{1\xi}\Lambda_{2\xi})^{-1/2} + (\Lambda_{1\eta}\Lambda_{2\eta})^{-1/2}\right)\operatorname{tr}\left(|\lambda(\boldsymbol{\Sigma}_\xi) - \lambda(\boldsymbol{\Sigma}_\eta)|\right),\end{aligned} \tag{8.5}$$

where $\Lambda_{k\xi}^2 := \sum_{j=k}^{\infty} \lambda_{j\xi}^2$, $\Lambda_{k\eta}^2 := \sum_{j=k}^{\infty} \lambda_{j\eta}^2$, $k = 1, 2$.

Note that the expression $\mathrm{tr}\big(|\lambda(\boldsymbol{\Sigma}_\xi) - \lambda(\boldsymbol{\Sigma}_\eta)|\big)$ on the right-hand side of (8.5) does not exceed $\|\boldsymbol{\Sigma}_\xi - \boldsymbol{\Sigma}_\eta\|_1$ (see, e.g., [7]). Additionally, the right-hand side of (8.5) involves $(\Lambda_{1\xi}\Lambda_{2\xi})^{-1/2}$ and $(\Lambda_{1\eta}\Lambda_{2\eta})^{-1/2}$), which appear in estimating the density functions of $\|\boldsymbol{\xi}\|^2$ and $\|\boldsymbol{\eta}\|^2$), respectively (see, e.g., formula (8.11)). Under various conditions on the spectrum of the covariance operator, in particular, taking into account the multiplicity of the largest eigenvalue, estimates for the density of the squared norm of a Gaussian vector were proved (see, e.g., [4]). In this chapter, however, we obtain an inequality with a more accurate dependence on the covariance matrix.

Theorem 8.1 can be extended to the non-centered balls (see Theorem 2.1 and its Corollaries in [6]):

Theorem 8.2 *Under conditions of Theorem 8.1 for all $a \in \mathbb{H}$ one has*

$$\sup_{x>0} |\Pr\{\|\boldsymbol{\xi} - \boldsymbol{a}\| \le x\} - \Pr\{\|\boldsymbol{\eta}\| \le x\}|$$
$$\lesssim \big((\Lambda_{1\xi}\Lambda_{2\xi})^{-1/2} + (\Lambda_{1\eta}\Lambda_{2\eta})^{-1/2}\big)\big(\|\boldsymbol{\Sigma}_\xi - \boldsymbol{\Sigma}_\eta\|_1 + \|\boldsymbol{a}\|^2\big),$$

Now we get some lower bounds to justify the structure of estimates in Theorems 8.1 and 8.2.

For simplicity, we consider the case of centerd ball, i.e., $\boldsymbol{a} = 0$. We show that in the case $\mathbb{H} = \mathbb{R}^2$ there exist covariance operators $\boldsymbol{\Sigma}_\xi$ and $\boldsymbol{\Sigma}_\eta$ such that

$$\sup_{x>0} |\Pr\{\|\boldsymbol{\xi}\| \le x\} - \Pr\{\|\boldsymbol{\eta}\| \le x\}| \gtrsim (\lambda_{1\xi}\lambda_{2\xi})^{-1/2}\, \|\boldsymbol{\Sigma}_\xi - \boldsymbol{\Sigma}_\eta\|_1, \tag{8.6}$$

i.e., in this case the lower bound coincides up to an absolute constant with the upper bound in Theorem 8.1. To show (8.6) we consider the following example. Let $\boldsymbol{\xi}$ and $\boldsymbol{\eta}$ be the Gaussian random vectors in $\mathbb{R}^2$ with zero means and covariance matrices $\boldsymbol{\Sigma}_\xi = \mathrm{diag}(\lambda_{1\xi}, \lambda_{2\xi})$ and $\boldsymbol{\Sigma}_\eta = \mathrm{diag}(\lambda_{1\eta}, \lambda_{2\eta})$ resp. Then

$$\sup_{x>0} |\Pr\{\|\boldsymbol{\xi}\| \le x\} - \Pr\{\|\boldsymbol{\eta}\| \le x\}| \ge \left|\Pr\{\|\boldsymbol{\xi}\| \le \sqrt{R}\} - \Pr\{\|\boldsymbol{\eta}\| \le \sqrt{R}\}\right|,$$

for some R which will be chosen later. Put

$$\mathcal{E}_1 \stackrel{\mathrm{def}}{=} \left\{(x_1, x_2) \in \mathbb{R}^2 : \sum_{j=1}^{2} \lambda_{j\xi} x_j^2 \le R\right\},$$
$$\mathcal{E}_2 \stackrel{\mathrm{def}}{=} \left\{(x_1, x_2) \in \mathbb{R}^2 : \sum_{j=1}^{2} \lambda_{j\eta} x_j^2 \le R\right\}.$$

Let us take $\lambda_{1\xi} = \lambda_{1\eta}$, $\lambda_{2\eta}/2 < \lambda_{2\xi} < \lambda_{2\eta}$. This choice gives $\|\boldsymbol{\Sigma}_\xi - \boldsymbol{\Sigma}_\eta\|_1 = \lambda_{2\eta} - \lambda_{2\xi}$ and $\Lambda_{1\xi}\Lambda_{2\xi} \asymp \Lambda_{1\eta}\Lambda_{2\eta} \asymp \lambda_{1\xi}\lambda_{2\xi}$, where $a \asymp b$ means that there exist the constants c, C such that $c\,a \le b \le C\,a$. It is straightforward to check that

$$\left|\Pr\{\|\boldsymbol{\xi}\| \le \sqrt{R}\} - \Pr\{\|\boldsymbol{\eta}\| \le \sqrt{R}\}\right| = \frac{1}{2\pi}\int_{\mathcal{E}_1\setminus\mathcal{E}_2} \exp\left(-\frac{x_1^2+x_2^2}{2}\right) dx_1\,dx_2$$
$$\ge \frac{1}{2\pi}(|\mathcal{E}_1| - |\mathcal{E}_2|)\exp\left[-\frac{R}{2}\left(\frac{1}{\lambda_{1\xi}} + \frac{1}{\lambda_{2\xi}}\right)\right],$$

where $|\mathcal{E}_i|$ is a volume of the ellipsoid $\mathcal{E}_i$, $i = 1, 2$. Applying the formula for the volume of an ellipsoid we obtain

$$|\mathcal{E}_1| - |\mathcal{E}_2| \ge \frac{\pi R\|\boldsymbol{\Sigma}_\xi - \boldsymbol{\Sigma}_\eta\|_1}{4\sqrt{2}\sqrt{\lambda_{1\xi}\lambda_{2\xi}\lambda_{2\xi}}}.$$

We take $R = 2\lambda_{2\xi}$. Then $R/(2\lambda_{2\xi})\exp\left(-R/(2\lambda_{2\xi})\right) = e^{-1} > 1/3$. Hence,

$$\left|\Pr\{\|\boldsymbol{\xi}\| \le \sqrt{R}\} - \Pr\{\|\boldsymbol{\eta}\| \le \sqrt{R}\}\right| \ge \frac{\|\boldsymbol{\Sigma}_\xi - \boldsymbol{\Sigma}_\eta\|_1}{12\sqrt{2}\sqrt{\lambda_{1\xi}\lambda_{2\xi}}}\exp\left[-\frac{\lambda_{2\xi}}{\lambda_{1\xi}}\right]$$
$$\gtrsim \frac{\|\boldsymbol{\Sigma}_\xi - \boldsymbol{\Sigma}_\eta\|_1}{\sqrt{\lambda_{1\xi}\lambda_{2\xi}}},$$

From the last inequality we get (8.6).

However it is still an open question to get a lower bound in Theorem 8.1 in general case.

8.4 Anti-concentration Inequality

It is well known that an arbitrary Gaussian element in $\mathbb{H}$ with a zero mean can be represented in the form

$$\boldsymbol{\xi} \overset{d}{=} \sum_{k=1}^{\infty} \sqrt{\lambda_{k\xi}}\, Z_k \mathbf{u}_k, \tag{8.7}$$

where Z_k, $k \ge 1$, are independent standard normal random variables; $\lambda_{k\xi}$, $k \ge 1$, are the eigenvalues of $\boldsymbol{\Sigma}_\xi$, arranged in nonincreasing order; and $\{\mathbf{u}_k\}_{k\ge1}$ are the orthonormal eigenvectors of $\boldsymbol{\Sigma}_\xi$ corresponding to the eigenvalues $\lambda_{k\xi}$. The following lemma provides an upper bound for the maximum of the density $p_\xi(x, \boldsymbol{a})$ of the random variable $\|\boldsymbol{\xi} - \boldsymbol{a}\|^2$ in terms of the eigenvalues of $\boldsymbol{\Sigma}_\xi$.

Lemma 8.1 *Let $\boldsymbol{\xi}$ be a Gaussian element in a separable Hilbert space $\mathbb{H}$ with zero mean and covariance operator $\boldsymbol{\Sigma}_\xi$. Then for all $\boldsymbol{a} \in \mathbb{H}$ one has*

$$\max_{x\ge0} p_\xi(x, \boldsymbol{a}) \lesssim (\Lambda_{1\xi}\Lambda_{2\xi})^{-1/2}. \tag{8.8}$$

In particular,

$$\max_{x \geq 0} p_{\xi}(x, \boldsymbol{a}) \lesssim (\lambda_{1\xi}\lambda_{2\xi})^{-1/2}. \tag{8.9}$$

Estimate (8.9) was proved in [13]. However, (8.8) is much more accurate. The following three typical situations can be distinguished. In the "one-dimensional" case with $\Lambda_{2\xi} \approx 0$, the assertion of the lemma is, in fact, meaningless; i.e., the effective dimension of the problem has to be at least two. In the "two-dimensional" case with $\Lambda_{2\xi} \approx \lambda_{2\xi}$, results (8.8) and (8.9) coincide. Finally, in the "multidimensional" case with $\lambda_{1\xi} \ll \Lambda_{1\xi}$, the quantities $\Lambda_{1\xi}$, $\Lambda_{2\xi}$ are of the same order and the right-hand side of (8.8) is inversely proportional to the Frobenius norm $\Lambda_{1\xi}$ of the operator $\boldsymbol{\Sigma}_{\xi}$. In particular, in the finite-dimensional case $\mathbb{H} = \mathbb{R}^p$ with $p \geq 2$, if $\boldsymbol{\Sigma}_{\xi}$ is close to the identity matrix, then, according to (8.8), one has $\max_{x \geq 0} p_{\xi}(x, 0) \leq c\, p^{-1/2}$. This estimate agrees with the maximum of the chi-square probability density function with p degrees of freedom.

However, a lower bound for $\sup_{x \geq 0} p_{\xi}(x, \boldsymbol{a})$ in the general case is still an open question. A possible extension Lemma 8.1 is a non-uniform upper bound for the p.d.f. of $\|\boldsymbol{\xi} - \boldsymbol{a}\|^2$. In this direction for any $\lambda > \lambda_{1\xi}$ we can prove that

$$p_{\xi}(x, \boldsymbol{a}) \leq \frac{\exp\{-(x^{1/2} - \|\boldsymbol{a}\|)^2/(2\lambda)\}}{2\sqrt{\lambda_{1\xi}\lambda_{2\xi}}} \prod_{j=3}^{\infty} (1 - \lambda_{j\xi}/\lambda)^{-1/2}; \tag{8.10}$$

see Lemma B.1 and remark after it in [6]. One can even get two-sided bounds for $p_{\xi}(x, \boldsymbol{a})$, but under additional conditions, see, e.g., [4]. It is still an open question whether it is possible to replace in (8.10) the $\lambda_{k\xi}$'s in the denominator by $\Lambda_{k\xi}$, $k = 1, 2$.

Lemma 8.1 has the following simple corollary which gives anti-concentration inequality. It provides an estimate for the probability of hitting a δ-strip.

Theorem 8.3 *Let $\boldsymbol{\xi}$ be a Gaussian element in $\mathbb{H}$ with zero mean and covariance operator $\boldsymbol{\Sigma}_{\xi}$. Let $\delta > 0$. Then for all $x \geq 0$ and $\boldsymbol{a} \in \mathbb{H}$, one has*

$$\Pr\{x < \|\boldsymbol{\xi} - \boldsymbol{a}\|^2 < x + \delta\} \lesssim (\Lambda_{1\xi}\Lambda_{2\xi})^{-1/2}\,\delta.$$

8.5 Proofs

For simplicity we give the proofs for Theorem 8.1 and Lemma 8.1 for the case $\boldsymbol{a} = 0$ only. See the general case and further details in [4, 6, 8].

8.5.1 Proof of Theorem 8.1

Without loss of generality, we assume that both covariance operators $\mathbf{\Sigma}_\xi$ and $\mathbf{\Sigma}_\eta$ have a diagonal form with nonincreasing eigenvalues. For every $s : 0 \le s \le 1$, we define a Gaussian element $Z(s)$ in $\mathbb{H}$ with zero mean and a diagonal covariance operator $\mathbf{V}(s)$:

$$\mathbf{V}(s) \overset{\text{def}}{=} s\mathbf{\Sigma}_\xi + (1-s)\mathbf{\Sigma}_\eta.$$

Let $\lambda_1(s) \ge \lambda_2(s) \ge \cdots$ denote the eigenvalues of the operator $\mathbf{V}(s)$. Define the resolvent operator $\mathbf{G}(t,s) \overset{\text{def}}{=} (\mathbf{I} - 2it\,\mathbf{V}(s))^{-1}$. Obviously, $\mathbf{G}(t,s)$ is also a diagonal operator and the characteristic function $f(t,s)$ of $\|Z(s)\|^2$ has the form

$$f(t,s) = \mathrm{E}\exp\{it\|Z(s)\|^2\} = \exp\left\{-\frac{1}{2}\mathrm{tr}\,\log\bigl(\mathbf{I} - 2it\mathbf{V}(s)\bigr)\right\}.$$

It is well known (see, e.g., Sect. 6.2, Ex. 10 in [5]) that, for a continuous distribution function $F(x)$ and its characteristic function $f(t)$, the inversion formula

$$F(x) = \frac{1}{2} + \frac{i}{2\pi}\lim_{T\to\infty} \mathrm{V.P.}\int_{|t|\le T} e^{-itx} f(t)\frac{dt}{t}.$$

holds, here V.P. means principal value of the integral, i.e.,

$$\mathrm{V.P.}\int_{|t|\le T} e^{-itx} f(t)\frac{dt}{t} = \lim_{\varepsilon\to 0+}\int_{\varepsilon\le|t|\le T} e^{-itx} f(t)\frac{dt}{t}.$$

Let $x > 0$ be fixed. Then

$$\begin{aligned}&\Pr\{\|\xi\|^2 < x\} - \Pr\{\|\eta\|^2 < x\}\\ &= \frac{i}{2\pi}\lim_{T\to\infty}\mathrm{V.P.}\int_{|t|\le T}\frac{f(t,1)-f(t,0)}{t}e^{-itx}\,dt.\end{aligned}$$

Using the Newton–Leibniz formula, we obtain

$$f(t,1) - f(t,0) = \int_0^1 \frac{\partial f(t,s)}{\partial s}\,ds = \int_0^1 if(t,s)\mathrm{tr}\bigl\{(\mathbf{\Sigma}_\xi - \mathbf{\Sigma}_\eta)\mathbf{G}(t,s)\bigr\}\,ds.$$

Changing the order of integration yields the equality

$$\begin{aligned}&\Pr\{\|\xi\|^2 < x\} - \Pr\{\|\eta\|^2 < x\}\\ &= -\frac{1}{2\pi}\int_0^1\int_{-\infty}^{\infty}\mathrm{tr}\,\bigl\{(\mathbf{\Sigma}_\xi - \mathbf{\Sigma}_\eta)\mathbf{G}(t,s)\bigr\}\,f(t,s)e^{-itx}\,dt\,ds.\end{aligned}$$

For a fixed s, consider the expression

$$\frac{1}{2\pi}\int_{-\infty}^{\infty}\mu_j(t,s)f(t,s)e^{-itx}\,dt,$$

where $\mu_j(t,s) := (1-2it\lambda_j(s))^{-1}$ are the eigenvalues of $\mathbf{G}(t,s)$. Let $\overline{Z}_j(s)$ be a random variable having an exponential distribution with parameter $1/(2\lambda_j(s))$ and independent of $Z(s)$. Then,

$$\mathrm{E}(e^{it\overline{Z}_j(s)}) = \mu_j(t,s).$$

Moreover, $\mu_j(t,s)f(t,s)$ is the characteristic function of $\overline{Z}_j(s) + \|Z(s)\|^2$. Let $p_j(x,s)$ be the density function corresponding to $\mu_j(t,s)f(t,s)$. Then

$$\frac{1}{2\pi}\int_{-\infty}^{\infty}\mu_j(t,s)f(t,s)e^{-itx}\,dt = p_j(x,s).$$

Let $\mathbf{P}(x,s)$ denote a diagonal operator with values $p_j(x,s)$ on the main diagonal. Then

$$\frac{1}{2\pi}\int_{-\infty}^{\infty}\mathrm{tr}\left\{(\boldsymbol{\Sigma}_\xi - \boldsymbol{\Sigma}_\eta)\mathbf{G}(t,s)\right\} f(t,s)e^{-itx}\,dt = \mathrm{tr}\left\{(\boldsymbol{\Sigma}_\xi - \boldsymbol{\Sigma}_\eta)\mathbf{P}(x,s)\right\}.$$

It is straightforward that the second term does not exceed

$$\|\boldsymbol{\Sigma}_\xi - \boldsymbol{\Sigma}_\eta\|_1 \cdot \|\mathbf{P}(x,s)\|, \tag{8.11}$$

so it remains to estimate $\|\mathbf{P}(x,s)\|$, i.e., each $p_j(x,s)$. However, for any j, the quantity $\max_x p_j(x,s)$ does not exceed $\max_x p(x,s)$, where $p(x,s)$ is the density function of $\|Z(s)\|^2$. Applying Lemma 8.1 and integrating the result with respect to s, we obtain the assertion of the theorem.

8.5.2 *Proof of Lemma 8.1*

For expository purposes, we consider the case of $\mathbb{H} = \mathbb{R}^{2p}$ and write λ_k instead of $\lambda_{k\xi}$. Without loss of generality, it may be assumed that $2\lambda_1^2 \le \sum_{k=1}^{p}\lambda_{2k-1}^2$, $2\lambda_2^2 \le \sum_{k=1}^{p}\lambda_{2k}^2$. Otherwise, we can apply the convolution formula; inequality (8.9) was proved in [13] exactly by this method and induction arguments. Put $p_\xi(x) = p_\xi(x,0)$. Using representation (8.7) and denoting by $f(t)$, $f_k(t)$, $k = 1,\dots,2p$ the characteristic functions of $\|\xi\|^2$, $\lambda_k\xi_k^2$, respectively, we obtain

$$p_\xi(x) \le \frac{1}{2\pi}\int_{-\infty}^{\infty} |f(t)|\,dt = \frac{1}{2\pi}\int_{-\infty}^{\infty} \prod_{k=1}^{p} |f_{2k-1}(t)| \prod_{k=1}^{p} |f_{2k}(t)|\,dt$$
$$\le \left[\frac{1}{2\pi}\int_{-\infty}^{\infty} \prod_{k=1}^{p} |f_{2k-1}(t)|^2\,dt\right]^{1/2} \left[\frac{1}{2\pi}\int_{-\infty}^{\infty} \prod_{k=1}^{p} |f_{2k}(t)|^2\,dt\right]^{1/2},$$

where the Cauchy–Schwarz inequality was applied at the last step. The Hölder inequality yields

$$p_\xi^2(x) \le \frac{1}{2\pi} \prod_{k=1}^{p} \| f_{2k-1}^2 \|_{p_{2k-1}} \prod_{k=1}^{p} \| f_{2k}^2 \|_{p_{2k}},$$

where $\sum_{k=1}^{p} p_{2k-1}^{-1} = 1$, $\sum_{k=1}^{p} p_{2k}^{-1} = 1$, $p_k \ge 2, k = 1, \ldots, 2p$, and $\|f\|_q^q = \int_{\mathbb{R}} |f(x)|^q\,dx, q \ge 1$. The exact values of p_k will be specified later. Using the inequality $(1+x)^a \ge 1 + ax, a \ge 1, x \ge 0$, we have

$$\| f_k^2 \|_{p_k}^{p_k} = \int_{-\infty}^{\infty} \frac{1}{(1+4\lambda_k^2 t^2)^{p_k/2}}\,dt \le \int_{-\infty}^{\infty} \frac{dt}{1+2p_k\lambda_k^2 t^2} \le \frac{\pi}{\lambda_k\sqrt{2p_k}}.$$

Now let $p_{2k-1}^{-1} := \lambda_{2k-1}^2 / \sum_{k=1}^{p} \lambda_{2k-1}^2$, $p_{2k}^{-1} := \lambda_{2k}^2 / \sum_{k=1}^{p} \lambda_{2k}^2$. Then, for all $k = 1, \ldots, p$,

$$\| f_{2k-1}^2 \|_{p_{2k-1}}^{p_{2k-1}} \le \frac{\pi}{\sqrt{2\sum_{k=1}^{p} \lambda_{2k-1}^2}}.$$

Similar inequalities hold for even indices, which leads to the required result.

Here we considered Gaussian random elements in a separable Hilbert space $\mathbb{H}$ with zero means. A possible extension of these results to Gaussian measures with different means and covariance operators can be found, for example, in [4, 6].

References

1. Chernozhukov, V., Chetverikov, D., & Kato, K. (2013). Gaussian approximations and multiplier bootstrap for maxima of sums of high-dimensional random vectors. *The Annals of Statistics*, *41*(6), 2786–2819.
2. Chernozhukov, V., Chetverikov, D., & Kato, K. (2015). Comparison and anti-concentration bounds for maxima of Gaussian random vectors. *Probability Theory and Related Fields*, *162*(1–2), 47–70.
3. Chernozhukov, V., Chetverikov, D., & Kato, K. (2017). Central limit theorems and bootstrap in high dimensions. *The Annals of Probability*, *45*(4), 2309–2352.
4. Christoph, G., Prokhorov, Y., & Ulyanov, V. (1996). On distribution of quadratic forms in Gaussian random variables. *Theory of Probability and its Applications*, *40*(2), 250–260.
5. Chung, K. (2001). *A course in probability theory* (3rd ed.). San Diego: Academic Press Inc.

6. Götze, G., Naumov, A., Spokoiny, V., & Ulyanov, V. (2019). Large ball probabilities, Gaussian comparison and anti-concentration. *Bernoulli*, *25*(4A), 2538–2563.
7. Markus, A. (1964). The eigen- and singular values of the sum and product of linear operators. *Russian Mathematical Surveys*, *19*, 91–120.
8. Naumov, A. A., Spokoiny, V. G., Tavyrikov, Yu E, & Ulyanov, V. V. (2018). Nonasymptotic estimates for the closeness of Gaussian measures on balls. *Doklady Mathematics*, *98*(2), 490–493.
9. Naumov, A., Spokoiny, V., & Ulyanov, V. (2019). Bootstrap confidence sets for spectral projectors of sample covariance. *Probability Theory and Related Fields*, *174*(3–4), 1091–1132.
10. Panov, M., & Spokoiny, V. (2015). Finite sample Bernstein-von Mises theorem for semiparametric problems. *Bayesian Analysis*, *10*, 665–710.
11. Spokoiny, V., & Zhilova, M. (2015). Bootstrap confidence sets under model misspecification. *The Annals of Statistics*, *43*(6), 2653–2675.
12. Tsybakov, A. (2008). *Introduction to nonparametric estimation*. New York: Springer.
13. Ulyanov, V.V. (1995). On Gaussian measure of balls in H. In *Frontiers in Pure and Applied Probability, Proceedings of the 4th Russian–Finnish Symposium on Probability Theory and Mathematical Statistics*. Moscow: TVP Science.

Chapter 9
Approximations for Statistics Based on Random Sample Sizes

Abstract In practice, we often encounter situations where a sample size is not defined in advance and can be random itself. It is known that the asymptotic properties of the statistics can be radically changed when the non-random sample size is replaced by a random value. In this chapter, we consider the second-order Chebyshev–Edgeworth type and Cornish–Fisher type expansions based on Student's t- and Laplace distributions and their quantiles for samples with random size of a special kind. This is accomplished using the general transfer theorem, which allows constructing asymptotic expansions for distributions of randomly normalized statistics from the distributions of the considered non-randomly normalized statistics and of the random size of the underlying sample.

9.1 Introduction

In classical problems of mathematical statistics, sample size is traditionally considered to be deterministic and plays a role as a known parameter which is, as a rule, infinitely increasing. There are many practical situations, where it is almost impossible to have a fixed sample size. They often occur when observations are collected in a fixed time span. For example, in reliability testing this is the number of failed devices, in medicine—the number of patients with a specific disease, in finance—the number of market transactions, in queueing theory—the number of customers entering a store, in insurance—the number of claims. All these numbers are random variables. The use of samples with random sample sizes has been steadily growing over the years. For an overview of statistical inferences with a random number of observations and some applications see, e.g., [15] and the references therein. Gnedenko [17] considered the asymptotic properties of the distributions of sample quantiles for samples of random size. In [25] unknown sample sizes are studied in medical research for analysis of one-way fixed effects ANOVA models to avoid false rejections. Esquível et al. [15] considered inference for the mean with known and unknown variance and inference for the variance in the normal model. Prediction intervals for the future

Y. Fujikoshi and V. V. Ulyanov, *Non-Asymptotic Analysis of Approximations for Multivariate Statistics*, JSS Research Series in Statistics,
https://doi.org/10.1007/978-981-13-2616-5_9

observations for generalized order statistics and confidence intervals for quantiles based on samples of random sizes are studied in [1, 2], respectively. They illustrated their results with real biometric data set, the duration of remission of leukemia patients treated by one drug.

9.2 Notation and Examples

We use the following notation: $\mathbb{R}$ as real numbers, $\mathbb{N} := \{1, 2, \ldots\}$ as positive integers and $\mathbf{I}_A(x)$ as an indicator function. Let $X, X_1, X_2, \ldots \in \mathbb{R}$ and $N_1, N_2, \ldots \in \mathbb{N}$ be random variables defined on the same probability space $(\Omega, \mathbf{A}, \Pr)$. In statistics, the random variables $X_1, X_2, \ldots, X_m$ are observations. Let N_n be a random size of the underlying sample, which depends on natural parameter $n \in \mathbb{N}$. We suppose that random variables $N_n \in \mathbb{N}$ with $n \in \mathbb{N}$ are independent of random variables $X_1, X_2, \ldots$.

Let $T_m := T_m(X_1, \ldots, X_m)$ be some statistic based on a sample of non-random size $m \in \mathbb{N}$. Define the random variable T_{N_n} for every $n \in \mathbb{N}$:

$$T_{N_n}(\omega) := T_{N_n(\omega)}\left(X_1(\omega), \ldots, X_{N_n(\omega)}\right), \quad \omega \in \Omega, \tag{9.1}$$

i.e., T_{N_n} is some statistic obtained from a random sample $X_1, X_2, \ldots, X_{N_n}$.

General asymptotic expansions for T_{N_n} are given in [6] applying corresponding asymptotic expansions for the normalized statistic T_m and the suitable scaled random sample size N_n.

Many models lead to random sums and random means

$$S_{N_n} = \sum\nolimits_{k=1}^{N_n} X_k \quad \text{and} \quad T_{N_n} = \frac{1}{N_n}\sum\nolimits_{k=1}^{N_n} X_k, \tag{9.2}$$

respectively. If N_n and X_1 have finite expectations, the Wald identity for random sums $\mathrm{E}(S_{N_n}) = \mathrm{E}(N_n)\mathrm{E}(X_1)$ is a powerful tool in statistical inference, particularly in sequential analysis, see, e.g., [33]. In [29] it is proved that asymptotic normality of the index N_n automatically implies asymptotic normality of the corresponding random sum S_{N_n}.

The randomness of the sample size may crucially change asymptotic properties of T_{N_n}, see, e.g., [17] or [18].

A fundamental introduction to asymptotic distributions of random sums is given in [13]. Moderate and large deviations are investigated in [14, 23]. Many applications when N_n is geometrically distributed are given in [22]. Bounds on the total variation distance between geometric random sum of independent, nonnegative, integer-valued random variables, and the geometric distribution are studied in Sect. 3 in [26].

It is worth to mention that a suitable scaled factor by S_{N_n} or T_{N_n} affects the type of limit distribution. In fact, consider random sum S_{N_n} given in (9.2). For the sake of convenience, let $X_1, X_2, \ldots$ be independent standard normal random variables and $N_n \in \mathbb{N}$ be geometrically distributed with $\mathrm{E}(N_n) = n$ and N_n be independent of $X_1, X_2, \ldots$. Then one has

$$\bullet \ \Pr\left\{\frac{1}{\sqrt{N_n}} S_{N_n} \le x\right\} = \int_{-\infty}^{x} \frac{1}{\sqrt{2\pi}} e^{-u^2/2} du \quad \text{for all } n \in \mathbb{N}, \tag{9.3}$$

$$\bullet \ \Pr\left\{\frac{1}{\sqrt{\mathrm{E}(N_n)}} S_{N_n} \le x\right\} \to \int_{-\infty}^{x} \frac{1}{\sqrt{2}} e^{-\sqrt{2}|u|} du \quad \text{as } n \to \infty, \tag{9.4}$$

$$\bullet \ \Pr\left\{\frac{\sqrt{\mathrm{E}(N_n)}}{N_n} S_{N_n} \le x\right\} \to \int_{-\infty}^{x} \left(2 + u^2\right)^{-3/2} du \quad \text{as } n \to \infty. \tag{9.5}$$

Thus, we have three different limit distributions. The suitable scaled geometric sum S_{N_n} is standard normal distributed or tends to the Laplace distribution with variance 1 depending on whether we take the random scaling factor $1/\sqrt{N_n}$ or the non-random scaling factor $1/\sqrt{\mathrm{E}N_n}$, respectively. Moreover, we get the Student distribution with 2 degrees of freedom as the limit distribution if we use scaling with the mixed factor $\sqrt{\mathrm{E}(N_n)}/N_n$. Similar results also hold for the normalized random mean T_{N_n}.

Assertion (9.3) we obtain by conditioning and by using the stability of the normal law. Moreover, by Stein's method, quantitative Berry–Esseen bounds in (9.3) and (9.4) for arbitrary centered random variables X_1 with $\mathrm{E}(|X_1|^3) < \infty$ were proved in Theorem 10.6 in [8], in Theorems 2.5 and 2.7 in [13] and in Theorem 3 in [28], respectively. Statement (9.5) follows from Theorem 2.1 in [3].

If the statistic T_m is asymptotically normal, then the limit laws of normalized statistic T_{N_n} are scale mixtures of normal distributions with zero mean, depending on the random sample size N_n and on the scaling factor.

Here we consider the second-order Chebyshev–Edgeworth type and Cornish–Fisher type expansions based on Student's t- and Laplace distributions and their quantiles for samples with particular random sizes. The importance and different practical applications of these limit distributions are discussed for example in [3, 4, 30]. Classical Cornish–Fisher expansions based on quantiles of the standard normal distribution were introduced in [12] their generalizations were proposed in [19]; see also [31]. Recently, interest in Cornish–Fisher expansions has increased because of studies in risk management. A widely known risk measure, Value at Risk (VaR), substantially depends on the quantiles of the loss function, which is connected with investment portfolio descriptions of financial instruments, for example, see [20].

In this chapter, we consider a sequence of independent identically distributed (i.i.d.) random variables $X, X_1, X_2, \ldots$ with

$$\left.\begin{aligned} &\mathrm{E}(|X|^5) < \infty, \ \mathrm{E}(X) = \mu, \ 0 < \mathrm{Var}(X) = \sigma^2, \\ &\lambda_3 = \sigma^{-3}\,\mathrm{E}((X-\mu)^3), \ \lambda_4 = \sigma^{-4}\,\mathrm{E}((X-\mu)^4) - 3 \end{aligned}\right\}. \tag{9.6}$$

Further, suppose that random variable X admits Cramér's condition:

$$\limsup_{|t|\to\infty} \left|\mathrm{E}(e^{itX})\right| < 1. \tag{9.7}$$

As statistic T_m, we consider the asymptotically normal sample mean:

$$T_m = (X_1 + \cdots + X_m)\,/m\,, \quad m = 1, 2, \ldots, \tag{9.8}$$

Then we have, based on Theorem 5.18 in [27] with $k = 5$),

$$\sup_x \left|\Pr\{\sigma^{-1}\sqrt{m}(T_m - \mu) \le x\} - \Phi_{2;m}(x)\right| \le Cm^{-3/2}, \tag{9.9}$$

where C does not depend on m and

$$\Phi_{2;m}(x) = \Phi(x) - \left\{\frac{\lambda_3}{6\sqrt{m}}H_2(x) + \frac{1}{m}\Big(\frac{\lambda_4}{24}H_3(x) + \frac{\lambda_3^2}{72}H_5(x)\Big)\right\}\varphi(x)$$

is the second-order asymptotic expansion with $\Phi(x)$ and $\varphi(x)$ as the distribution function and density function of the standard normal random variable, respectively, and

$$H_2(x) = x^2 - 1\,, \quad H_3(x) = x^3 - 3x \quad \text{and} \quad H_5(x) = x^5 - 10x^3 + 15x$$

are the Chebyshev–Hermite polynomials.

Consider now the random mean T_{N_n}, based on statistic (9.8), see (9.8) Let g_n be a sequence of positive real numbers: $0 < g_n \uparrow \infty$. Suppose that $N_n \to \infty$ in probability as $n \to \infty$.

The limit laws of $\Pr\left\{\sigma^{-1}\sqrt{g_n}(T_{N_n} - \mu) \le x\right\}$ are scale mixtures of normal distributions with zero mean, depending on N_n. In Sect. 9.2, we provide auxiliary statements to find Chebyshev–Edgeworth and Cornish–Fisher expansions for the normalized random mean T_{N_n}. In Sect. 9.3, as random sample size N_n, we consider the negative binomial distribution, shifted by 1, with success probability $p = 1/n$. This is one of the leading distributions for count models. If we take $g_n = \mathrm{E}(N_n)$, then N_n/g_n tends to the Gamma distribution and the normalized random mean T_{N_n} converges to Student's t-distribution. In Sect. 9.4, the random size N_n is the maximum of n i.i.d. discrete Pareto random variables with tail parameter 1, where N_n/g_n with $g_n = n$ tends to the reciprocal exponential distribution. In this case, the Laplace law is the limit distribution of the normalized random mean T_{N_n}. In both cases, we need the second Chebyshev–Edgeworth-type expansions for N_n/g_n. The details of proofs see in [11].

Similar results for sample median in the case of random size samples see in [10].

9.3 Two Transfer Propositions

We suppose that the statistic $T_m = T_m(X_1, \ldots, X_m)$ satisfies the following condition when a sample size $m \in \mathbb{N}$ is non-random.

Condition 9.1 *There exist differentiable bounded functions $f_1(x)$, $f_2(x)$ and real numbers $a > 1$, $C_1 > 0$ such that, for all integers $m \geq 1$,*

$$\sup_x \left| \Pr\left\{\sigma^{-1}\sqrt{m}(T_m - \mu) \leq x\right\} - \Phi(x) - m^{-1/2} f_1(x) - m^{-1} f_2(x) \right| \leq C_1 m^{-a}. \tag{9.10}$$

Consider now the statistic $T_{N_n} = T_{N_n}\left(X_1, \ldots, X_{N_n}\right)$ with a random number $N_n = N_n \in \mathbb{N}$ of observations satisfying condition.

Condition 9.2 *There exists a distribution function $H(y)$ with $H(0+) = 0$, a function of bounded variation $h_2(y)$, a sequence $0 < g_n \uparrow \infty$ and real numbers $b > 1$ and $C_2 > 0$ such that, for all integers $n \geq 1$,*

$$\sup_{y \geq 0} \left| \Pr\{g_n^{-1} N_n \leq y\} - H(y) - n^{-1} h_2(y) \right| \leq C_2 n^{-b}. \tag{9.11}$$

Proposition 9.1 *Let $X, X_1, X_2, \ldots$ be i.i.d. random variables satisfying* (9.6) *and Cramér's condition* (9.7)*, i.e., due to* (9.9) *Condition 9.1 holds and* (9.10) *is satisfied with $a = 3/2$ and*

$$f_1(x) = -\frac{\lambda_3}{6} H_2(x)\varphi(x), \quad f_2(x) = -\left\{\frac{\lambda_4}{24} H_3(x) + \frac{\lambda_3^2}{72} H_5(x)\right\}\varphi(x).$$

Suppose Condition 9.2 with (9.11) *holds for N_n with the additional assumptions for some $\gamma > 1$: $h_2(0) = 0$, $H(g_n^{-1}) \leq c_0 n^{-\gamma}$ and $h_2(g_n^{-1}) \leq c_1 n^{1-\gamma}$. Then there exists a constant $C_3 = C_3(\lambda_3, \lambda_4, C_2) > 0$ such that, $\forall n \in \mathbb{N}$,*

$$\sup_x \left| \Pr\{\sigma^{-1}\sqrt{g_n}(T_{N_n} - \mu) \leq x\} - G_{2,n}(x) \right| \leq C_1 \mathrm{E}(N_n^{-a}) + C_3 n^{-\min\{b,\gamma\}},$$

where

$$G_{2;n}(x) = \int_0^\infty \Phi(x\sqrt{y})dH(y) + \frac{1}{\sqrt{g_n}} \int_0^\infty \frac{f_1(x\sqrt{y})}{\sqrt{y}} dH(y)$$
$$+ \frac{1}{g_n} \int_0^\infty \frac{f_2(x\sqrt{y})}{y} dH(y) + \frac{1}{n} \int_0^\infty \Phi(x\sqrt{y}) dh_2(y).$$

Suppose now that $F_n(x)$ is a sequence of distribution functions admitting a Chebyshev–Edgeworth expansion in powers of $g_n^{-1/2}$ with $0 < g_n \uparrow \infty$ as $n \to \infty$, that is,

$$F_n(x) = G(x) + g(x)\big(a_1(x)g_n^{-1/2} + a_2(x)g_n^{-1}\big) + R(g_n),$$
$$R(g_n) = o(g_n^{-1}), \text{ as } n \to \infty, \tag{9.12}$$

where $g(x)$ is a density function of the limit distribution $G(x)$.

Proposition 9.2 *Let $F_n(x)$ satisfy* (9.12) *and let $x(u)$ and u be quantiles of distributions F_n and G of the same order, that is $F_n(x(u)) = G(u)$. Then one has the following expansion:*

$$x(u) = u + b_1(u)g_n^{-1/2} + b_2(u)g_n^{-1} + R^*(g_n), \quad R^*(g_n) = o(g_n^{-1}), \quad n \to \infty,$$

with

$$b_1(u) = -a_1(u) \quad \textit{and} \quad b_2(u) = \frac{g'(u)}{2\,g(u)}a_1^2(u) + a_1'(u)a_1(u) - a_2(u)\,.$$

Proposition 9.2 is a direct consequence of the more general statements in Sect. 5.6.1 in [16] and in [32].

9.4 Edgeworth and Cornish–Fisher Expansions with Student's Limit Distribution

Student's t-distribution function $S_\nu(x)$ is an absolutely continuous probability distribution function given by the density

$$s_\nu(x) = \frac{\Gamma((\nu+1)/2)}{\sqrt{\nu\pi}\,\Gamma(\nu/2)}\left(1 + \frac{x^2}{\nu}\right)^{-(\nu+1)/2}, \quad x \in \mathbb{R}, \tag{9.13}$$

where $\nu > 0$ is a real shape parameter. If the parameter ν is a positive integer, then it is called the number of degrees of freedom. Student's t-distributions are the limit laws for T_{N_n} defined in (9.1), when $T_m := T_m(X_1, \ldots, X_m)$ is asymptotically normal, the r.v. $X_1, X_2, \ldots$ have finite variances and the sample sizes N_n have a negative binomial distribution, independent of $X_1, X_2, \ldots$ and $N_n \to \infty$ in probability as $n \to \infty$.

Recall that $N_n(r)$ is negative binomial distributed (shifted by 1) with success probability $1/n$ when its probability mass function equals

$$\Pr\{N_n(r) = j\} = \frac{\Gamma(j + r - 1)}{(j-1)!\,\Gamma(r)}\left(\frac{1}{n}\right)^r\left(1 - \frac{1}{n}\right)^{j-1}, j = 1, 2, \ldots \; r > 0. \tag{9.14}$$

Since $N_n(r)$ takes values $\{1, 2, \ldots\}$, the random mean $T_{N_n(r)}$ is well defined. If r is a positive integer, then $r \geq 1$ is the predefined number of successes and $N_n(r) - 1$ is the random number of failures until the experiment is stopped. Moreover, for fixed $k \in \mathbb{N}$ (see formula (5.31) on p. 218 in [21]) we have the following expression for the distribution function:

$$\Pr\{N_n(r) \leq k\} = \sum\nolimits_{j=1}^{k} \frac{\Gamma(j+r-1)}{(j-1)!\,\Gamma(r)} \left(\frac{1}{n}\right)^r \left(1 - \frac{1}{n}\right)^{j-1} = \frac{B_{1/n}(r,k)}{B(r,k)} \tag{9.15}$$

with beta function $B(r, k) = \Gamma(k)\,\Gamma(r)/\Gamma(k + r)$ and incomplete beta function

$$B_{1/n}(r,k) = \int_0^{1/n} u^{r-1}\,(1-u)^{k-1} du \overset{u=t/(1+t)}{=} \int_0^{1/(n-1)} \frac{t^{r-1}}{(1+t)^{k+r}} dt.$$

Set $g_n := \mathrm{E}(N_n(r)) = r\,(n-1) + 1$; then

$$\sup\nolimits_{x>0} \left|\Pr\{N_n(r)/\mathrm{E}(N_n(r)) \leq x\} - G_{r,r}(x)\right| \to 0 \quad \text{as} \quad n \to \infty,$$

where $G_{\alpha,\beta}(x)$ is the gamma distribution function with the shape parameter $\alpha > 0$ and rate parameter $\beta > 0$, having density

$$g_{\alpha,\beta}(x) = \frac{\beta^\alpha}{\Gamma(\alpha)} x^{\alpha-1} e^{-\beta x}\, \mathbf{I}_{(0\,,\,\infty)}(x), \quad x \in \mathbb{R}.$$

For an asymptotically normal random mean T_m, the limit distribution of $\Pr\left\{\sigma^{-1}\sqrt{r(n-1)+1}(T_{N_n(r)} - \mu) \leq x\right\}$ is Student's t-distribution $S_{2r}(x)$ with density function (9.13) and shape parameter $\nu = 2r$; see Corollary 2.1 in [3] or Theorem 1 in [30]. In [5, 6], under some moment conditions, proofs of the rates of convergence and first-order Chebyshev–Edgeworth-type expansions of asymptotically normal statistics T_m based on samples with random size $N_n(r)$ are provided. In order to obtain second-order Chebyshev–Edgeworth expansion for the normalized $T_{N_n(r)}$, we have to prove a second-order Chebyshev–Edgeworth expansion for $N_n(r)/\mathrm{E}(N_n(r))$; see (9.11) in Condition 9.2 above. Since the limit Gamma distribution $G_{r,r}(x)$ is a continuous function and $\Pr\{N_n(r)/\mathrm{E}(N_n(r)) \leq x\}$ is a step function, it is necessary to add a discontinuous function to overcome the jumps.

Theorem 9.1 *Let $r > 1$ and the discrete random variable $N_n(r)$ have probability mass function* (9.14) *and* $\mathrm{E}(N_n(r)) = r(n-1) + 1$. *For $x > 0$ and all $n \in \mathbb{N}$, there exists a real number $C_2(r) > 0$ such that*

$$\sup\nolimits_{x\geq 0} \left| \Pr\left\{\frac{N_n(r)}{r(n-1)+1} \leq x\right\} - G_{2;n}(x)\right| \leq C_2(r)\, n^{-\min\{r,2\}}, \tag{9.16}$$

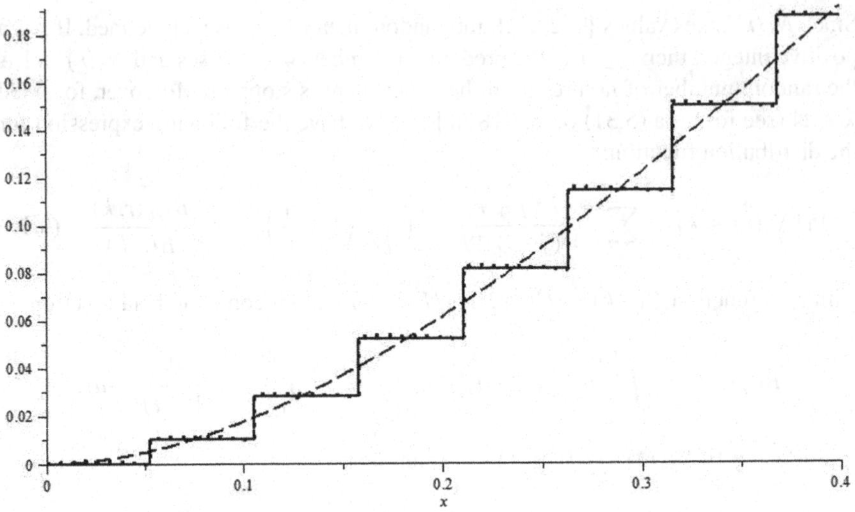

Fig. 9.1 Distribution function $\Pr(N_n(r) \le (r(n-1)+1)x)$ (solid line), the limit law $G_{2,2}(x)$ (dashed line) and the second-order approximation $G_{2,2}(x) + h_2(x)/n$ (doted line) for $n = 10$ and $r = 2$

where

$$G_{2;n}(x) = G_{r,r}(x) + \frac{a_0 G_{r,r}(x) + a_1 G_{r+1,r}(x) + a_2 G_{r+2,r}(x)}{2\,x\,n} \mathbf{I}_{[2,\infty)}(n) \tag{9.17}$$

$$= G_{r,r}(x) + \frac{g_{r,r}(x)\left((x-1)(2-r) + 2Q_1\big((r(n-1)+1)x\big)\right)}{2\,r\,n} \mathbf{I}_{[2,\infty)}(n), \tag{9.18}$$

$$a_1 = -2(x+r) + 1 - 2Q_1\big((r(n-1)+1)x\big),\ a_2 = r+1,\ a_0 = -(a_1+a_2),$$
$$Q_1(y) = 1/2 - (y - [y]), \tag{9.19}$$

where $[y]$ *is the integer part of* y *with* $y - 1 < [y] \le y$.

Remark 9.1 Formula (9.19) shows that (9.11) is satisfied with $H(x) = G_{r,r}(x)$, $h_2(x) = \big((x-1)(2-r) + 2Q_1\big(r(n-1)+1)x\big)\big) g_{r,r}(x)\mathbf{I}_{[2,\infty)}(n)/(2r)$ for $x > 0$, $b = \min\{r, 2\}$, and $g_n = r(n-1)+1$.

Figure 9.1 shows the approximation of $\Pr(N_n(r) \le (r(n-1)+1)x)$ by $G_{2,2}(x)$ and $G_{2,2}(x) + h_2(x)/n$ for $n = 10$ and $r = 2$, see also [11].

Remark 9.2 Coefficients a_0, a_1, a_2 in (9.17) meet condition $a_0 + a_1 + a_2 = 0$. In [32] it is considered a wide class of statistics allowing representation similar to (9.17).

Remark 9.3 Integration by parts allows passage from (9.17) to (9.19) because

$$G_{r+1,r}(x) = -(x/r)g_{r,r}(x) + G_{r,r}(x) \tag{9.20}$$

and

$$G_{r+2,r}(x) = -\big(x^2/(r+1) + x/r\big)g_{r,r}(x) + G_{r,r}(x).$$

In the following case, it is straightforward to prove the bound (9.16).

Remark 9.4 If $n = 1$, then $\Pr\{N_1(r) = 1\} = 1$ and the left-hand side in (9.16) takes its maximum at $x = 1$, leading to $C_2(r) = \min\{G_{r,r}(1), 1 - G_{r,r}(1)\} < 1$.

In addition to the expansion of $N_n(r)$, a bound of $\mathrm{E}(N_n(r)^{-a})$ is required, where m^{-a} is the rate of convergence of the Chebyshev–Edgeworth expansion for T_m; see (9.10).

Lemma 9.1 *Let $r > 1$ and the random variable $N_n(r)$ be defined by* (9.14). *Then*

$$\mathrm{E}\big(N_n(r)^{-3/2}\big) \le C(r)\begin{cases} n^{-r}, & 1 < r < 3/2, \\ \ln(n)\, n^{-3/2}, & r = 3/2, \\ n^{-3/2}, & r > 3/2. \end{cases} \tag{9.21}$$

In case of $r = 3/2$, the convergence rate in (9.21) *cannot be improved.*

Now we present the second-order Chebyshev–Edgeworth expansion for the standardized random mean $T_{N_n(r)}$.

Theorem 9.2 *Let $X, X_1, X_2, \ldots$ be i.i.d. random variables, where X satisfies* (9.6) *and Cramér's condition* (9.7). *Let the discrete random variable $N_n = N_n(r)$ with parameter $r > 1$ have probability mass function* (9.14), *being independent of $X_1, X_2, \ldots$. Consider the statistic $T_{N_n} = N_n^{-1}(X_1 + \cdots + X_{N_n})$. Suppose that for the asymptotically normal statistics T_m, the asymptotic expansion* (9.9) *holds and for the random size $N_n(r)$ with $r > 1$, one has the asymptotic expansion* (9.16) *with $g_n = \mathrm{E}\big(N_n(r)\big) = r(n-1) + 1$. Then there exists a constant $C = C(r) > 0$ such that, for all $n \in \mathbb{N}$,*

$$\sup_x \Big| \Pr\Big\{\sigma^{-1}\sqrt{g_n}(T_{N_n} - \mu) \le x\Big\} - S_{2r;n}(x)\Big| \le C\begin{cases} n^{-r}, & 1 < r < 3/2, \\ \ln(n)\, n^{-3/2}, & r = 3/2, \\ n^{-3/2}, & r > 3/2, \end{cases} \tag{9.22}$$

where

$$\begin{aligned} S_{2r;n}(x) = S_{2r}(x) &- \frac{\lambda_3\,\big((r-1)x^2 - r\big)\,s_{2r}(x)}{3\,(2r-1)\sqrt{g_n}} \\ &- \frac{x\,s_{2r}(x)}{36\,(2r-1)\,g_n}\left\{\frac{2\,\lambda_3^2\,\big((r-2)(r-3)x^4 + 10r\,(2-r)x^2 + 15r^2\big)}{2r + x^2}\right. \\ &\left. + 3\,\lambda_4\,\big((r-2)x^2 - 3r\big) + 9(r-2)(x^2+1)\right\}. \end{aligned}$$

In the case $r = 3/2$, the convergence rate in (9.22) cannot be improved.

Using the second-order Chebyshev–Edgeworth expansion in Theorem 9.2 and transfer Proposition 9.2, we obtain the following statement.

Theorem 9.3 *Let $x = x_\alpha$ and $u = u_\alpha$ be α-quantiles of standardized statistic $\sigma^{-1}\sqrt{r(n-1)+1}\left(T_{N_n(r)} - \mu\right)$ and of the limit Student's t-distribution $S_{2r}(x)$, respectively. Then for $u \neq 0$, the following asymptotic expansion holds as $n \to \infty$:*

$$x = u + \frac{\lambda_3\left((r-1)u^2 - r\right)}{3\,(2r-1)\sqrt{g_n}} + \frac{1}{g_n}B_2(u) + \begin{cases} O(n^{-r}), & 1 < r < 3/2, \\ O(\ln(n)\, n^{-3/2}), & r = 3/2, \\ O(n^{-3/2}), & r > 3/2, \end{cases}$$

where

$$B_2(u) = -\frac{(2r+1)u}{2(2r+u^2)} \frac{\lambda_3^2\left((r-1)u^2 - r\right)^2}{9\,(2r-1)^2} + \frac{2\,\lambda_3^2(r-1)u\left((r-1)u^2 - r\right)}{9\,(2r-1)^2}$$
$$+ \frac{u}{36\,(2r-1)}\left\{\frac{2\,\lambda_3^2\left((r-2)(r-3)u^4 + 10\,r\,(2-r)u^2 + 15r^2\right)}{2r+u^2}\right.$$
$$\left. + 3\,\lambda_4\left((r-2)u^2 - 3r\right) + 9(r-2)(u^2+1)\right\}.$$

Remark 9.5 Since

$$g_n = r(n-1) + 1 \quad \text{and} \quad |g_n^{-\gamma} - (rn)^{-\gamma}| \leq C(r) n^{-\gamma-1},$$

the factors $g_n^{-1/2}$ and g_n^{-1} in both Chebyshev–Edgeworth and Cornish–Fisher expansions may be replaced by $(rn)^{-1/2}$ and $(rn)^{-1}$, respectively.

9.5 Edgeworth and Cornish–Fisher Expansions with Laplace Limit Distribution

Let $Y(s)$ be a discrete Pareto II distributed random variable with parameter $s > 0$ and probability mass function

$$\Pr\{Y(s) = k\} = \frac{s}{s+k-1} - \frac{s}{s+k}, \quad k \in \mathbb{N} = \{1, 2, \ldots\}, \quad s > 0, \tag{9.23}$$

which is a particular class of a more general model of discrete Pareto distributions, obtained by discretization of the continuous Pareto II (Lomax) distributions on integers; see, e.g., [7], where the main properties and characteristics of discrete Pareto distributions are discussed.

Let $s \geq 1$ be an integer, $Z_1, Z_2, \ldots$ be i.i.d. random variables with some continuous distribution function and

$$Y(s) = \min\{k \geq 1 : \max_{1 \leq j \leq s} Z_j < \max_{s+1 \leq j \leq s+k} Z_j\}.$$

Then $Y(s)$ has probability mass function (9.23). The random variable $Y(s)$ plays an important role in the theory of records and describes the number of trials required to obtain the next extreme observation; see, e.g., [34] or [4]. The telescoping nature of the probabilities in (9.23) leads to relations:

$$\Pr\left\{Y(s) \leq k\right\} = \sum_{m=1}^{k} \Pr\left\{Y(s) = m\right\} = 1 - \frac{s}{s+k} = \frac{k}{s+k}, \ k \in \mathbb{N}, \ s > 0. \tag{9.24}$$

Now, let $Y_1(s), Y_2(s), \ldots, s > 0$ be i.i.d. random variables with the same distribution (9.24). Define the random variables $N_n(s) = \max_{1 \leq j \leq n} Y_j(s)$. Using

$$\Pr\{N_n(s) = k\} = \Pr\{N_n(s) \leq k\} - \Pr\{N_n(s) \leq k-1\},$$

we find

$$\Pr\{N_n(s) = k\} = \left(\frac{k}{s+k}\right)^n - \left(\frac{k-1}{s+k-1}\right)^n, \ k \in \mathbb{N}, \ s > 0. \tag{9.25}$$

Now consider a random mean $T_{N_n(s)}$ given in (9.2), where $N_n(s)$ has probability mass function (9.25). Bening and Korolev in [4] proved for integer $s \geq 1$ that

$$\lim_{n\to\infty} \Pr\left\{N_n(s) \leq n\,x\right\} = H(x) = e^{-s/x}\, \mathbf{I}_{(0,\infty)}(x), \tag{9.26}$$

and the limit distribution of $\Pr\left\{\sigma^{-1}\sqrt{n}(T_{N_n(s)} - \mu) \leq x\right\}$ is the Laplace distribution function $L_{1/\sqrt{s}}(x)$ having density function $l_{1/\sqrt{s}}(x) = \sqrt{s/2}\,\exp\{-\sqrt{2\,s}|x|\}$, $x \in \mathbb{R}$. In [5, 6], rates of convergence in (9.26) and a first-order asymptotic expansion for $\Pr\left\{\sigma^{-1}\sqrt{n}(T_{N_n}(s) - \mu) \leq x\right\}$ are proved for integer $s \geq 1$.

Now we get the Chebyshev–Edgeworth expansion for $\Pr\left\{N_n(s)/n \leq x\right\}$ satisfying (9.11).

Theorem 9.4 *Let* $N_n = N_n(s)$ *have probability mass function* (9.25) *and* $Q_1(x)$ *be defined in* (9.19). *Then for* $x > 0$, *fixed* $s \geq s_0 > 0$ *and all* $n \in \mathbb{N}$, *there exists a real number* $C_2(s) > 0$ *such that*

$$\sup_{x>0}\left|\Pr\left\{\frac{N_n(s)}{n} \leq x\right\} - e^{-s/x}\left\{1 + \frac{s\left(s - 1 + 2Q_1(n\,x)\right)}{2\,x^2\,n}\right\}\right| \leq \frac{C_2(s)}{n^2}. \tag{9.27}$$

Remark 9.6 Formula (9.27) shows that (9.11) is satisfied with $g_n = n$,

$$H(x) = e^{-s/x}, \quad h_2(x) = e^{-s/x}\, s\left(s - 1 + 2Q_1(n\,x)\right) / (2\,x^2)$$

and $\beta = 2$.

Remark 9.7 In [24] it is proved a first-order bound in (9.27) for integer $s \geq 1$:

$$\left|\Pr\left\{\frac{N_n(s)}{n} \leq x\right\} - e^{-s/x}\right| \leq \frac{C(s)}{n}, \quad C(s) = \begin{cases} 8e^{-2}/3, & s = 1, n \geq 2, \\ 2e^{-2}, & s \geq 2, n \geq 1. \end{cases}$$

In the case $n = 1$ and $s = 1$, we have $\Pr\{N_1(1) \leq x\} = 0$ for $0 < x < 1$ and

$$\sup_{0<x<1}\left|\Pr\{N_1(1) \leq x\} - e^{-1/x}\right| = \sup_{0<x<1} e^{-1/x} = e^{-1} = 0.367\ldots.$$

Remark 9.8 The random variable $N_n(s)$ is discrete with integer values $k \geq 1$. Therefore, the distribution function $\Pr\{N_n(s) \leq n\,x\}$ is discontinuous with discontinuity points $x = k/n, k = 1, 2, \ldots$, whereas the limit distribution $H(x) = e^{-s/x}\,\mathbf{I}_{(0,\infty)}(x)$ is continuous. In the interval $(x, x + 1/n]$ with $x > 0$, the distribution function $\Pr\{N_n(s) \leq n\,x\}$ has only one jump point, $([nx] + 1)/n$. The increase of limit distribution $H(x)$ over the interval $(x, x + 1/n]$ is

$$H(x + 1/n) - H(x) = s\,e^{-s/x}/(n\,x^2) + \mathrm{O}(n^{-2}), \quad n \to \infty,$$

which is equivalent to the jump at $([nx] + 1)/n$ of the discontinuity correcting function in (9.27).

Remark 9.9 The continuous function $e^{-s/x}\mathbf{I}_{(0,\infty)}(x), s > 0$, is the distribution function of the reciprocal random variable $W(s) = 1/V(s)$, where $V(s)$ is exponentially distributed with the rate parameter $s > 0$.

In Theorems 9.2 and 9.3, the normalizing sequence is $g_n = \mathrm{E}\big(N_n(r)\big)$ when the random size has a negative binomial distribution; see (9.15). In Theorems 9.4 and 9.5, the normalizing sequence is $g_n = n$, since both $\Pr\big\{N_n(s) \leq n\,x\big\}$ and $e^{-s/x}\mathbf{I}_{(0,\infty)}(x)$ for fixed $s > 0$ are heavy tailed with shape parameter 1.

Lemma 9.2 *For random size $N_n(s)$ with probabilities* (9.25) *and random variable $W(s)$ with distribution function $e^{-s/x}\mathbf{I}_{(0,\infty)}(x)$, $s > 0$ and $1 \leq r < 2$, we have*

(i) $\mathrm{E}\big(N_n(s)\big) = \infty$ *and* $\mathrm{E}\big(W(s)\big) = \infty$,
(ii) pseudo moment $\nu_1 = \int_0^\infty x\big|d\big(\Pr\{N_n(s) \leq n\,x\} - e^{-s/x}\big)\big| = \infty$,
(iii) difference moment $\chi_r = \int_0^\infty x^{r-1}\big|\Pr\big\{N_n(s) \leq n\,x\big\} - e^{-s/x}\big|dx < \infty$.

For information on pseudo moments and some of their generalizations, see, e.g., Chap. 2 in [9].

In addition to the expansion of $N_n(s)$, a bound of $\mathrm{E}(N_n(s))^{-3/2}$ is required.

Lemma 9.3 *For random size $N_n(s)$ with probabilities (9.25), where $s \geq s_0 > 0$, for arbitrary small $s_0 > 0$, and $n \geq 1$, we have*

$$\mathrm{E}\big(N_n^{-3/2}(s)\big) \leq \left\{ \begin{matrix} \zeta(3/2) = 2.612\ldots & n = 1 \\ \dfrac{sn}{\Gamma(5/2)(n-3/2)^{5/2}}, & n \geq 2 \end{matrix} \right\} \leq C(s)n^{-3/2}.$$

Now we present the second-order Chebyshev–Edgeworth expansion for the standardized random mean $T_{N_n(s)}$.

Theorem 9.5 *Let $X, X_1, X_2, \ldots$ be i.i.d. random variables, where X satisfies (9.6) and Cramér's condition (9.7). Let the discrete random variable $N_n = N_n(s)$ with real parameter $s \geq s_0 > 0$ have probability mass function (9.25), being independent of $X_1, X_2, \ldots$. Consider $T_{N_n} = N_n^{-1}(X_1 + \cdots + X_{N_n})$. Suppose that, for T_m, the asymptotic expansion (9.9) holds and, for $N_n(s)$ with $s \geq s_0 > 0$, one has the asymptotic expansion (9.27) with $g_n = n$. Then there exists a constant $C(s) > 0$ such that, for all $n \in \mathbb{N}$, one has*

$$\sup_x \left|\Pr\left\{\sigma^{-1}\sqrt{n}(T_{N_n(s)} - \mu) \leq x\right\} - L_{1/\sqrt{s};n}(x)\right| \leq C(s)n^{-3/2},$$

where

$$L_{1/\sqrt{s};n}(x) = L_{1/\sqrt{s}}(x) + n^{-1/2}A_1(x)\,l_{1/\sqrt{s}}(x) \\ +n^{-1}\Big(A_{2,1}(x) + A_{2,2}(x)\Big)l_{1/\sqrt{s}}(x),$$

$$A_1(x) = \frac{\lambda_3}{6}\Big(-x^2 + \frac{|x|}{\sqrt{2s}} + \frac{1}{2s}\Big), \quad A_{2,2}(x) = \frac{x(1-s)}{8s}\,(\sqrt{2s}\,|x| + 1),$$

$$A_{2,1}(x) = \frac{x\,\lambda_4}{48\,s}\Big(3 - 2sx^2 + 3\sqrt{2s}|x|\Big) \\ +\frac{x\,\lambda_3^2}{144\,s}\Big(20sx^2 - (2s)^{3/2}|x|^3 - 15\sqrt{2s}|x| - 15\Big).$$

Using the second-order Chebyshev–Edgeworth expansion in Theorem 9.4 and transfer Proposition 9.2, we obtain the following statement.

Theorem 9.6 *Let $x = x_\alpha$ be the α-quantile of $\sigma^{-1}\sqrt{n}\left(T_{N_n(s)} - \mu\right)$ and let $u = u_\alpha$ be the α-quantile of the limit Laplace distribution $L_{1/\sqrt{s}}(x)$. Then for $u \neq 0$, the following asymptotic expansion holds:*

$$x = u - \frac{\lambda_3}{6\sqrt{n}}\Big(\frac{|u|}{\sqrt{2s}} + \frac{1}{2s} - u^2\Big) + \frac{1}{n}B_2(u) + \mathrm{O}\Big(n^{-3/2}\Big), \quad \textit{as } n \to \infty,$$

where

$$B_2(u) = -\frac{\lambda_3^2}{36}\left(\frac{\sqrt{2s}\,u}{2|u|}\left(u^2 - \frac{|u|}{\sqrt{2s}} - \frac{1}{2s}\right)^2\right.$$
$$\left. + \left(u^2 - \frac{|u|}{\sqrt{2s}} - \frac{1}{2s}\right)\left(2u - \frac{u}{\sqrt{2s}|u|}\right)\right)$$
$$+ \frac{u\,\lambda_3^2}{144\,s}\left(20su^2 - (2s)^{3/2}|u|^3 - 15\sqrt{2s}|u| - 15\right)$$
$$+ \frac{u\,\lambda_4}{48\,s}\left(3 - 2su^2 + 3\sqrt{2s}|u|\right) + \frac{u(1-s)}{8s}(\sqrt{2\,s}\,|u| + 1).$$

For the details of proofs, see [11].
For similar results related to sample median for random size samples, see [10].

References

1. Al-Mutairi, J. S., & Raqab, M. Z. (2017). Confidence intervals for quantiles based on samples of random sizes. Statistical Papers.
2. Barakat, H. M., Nigm, E. M., El-Adll, M. E., & Yusuf, M. (2018). Prediction of future generalized order statistics based on exponential distribution with random sample size. *Statistical Papers*, *59*, 605–631.
3. Bening, V. E., & Korolev, V Yu. (2005). On an application of the Student distribution in the theory of probability and mathematical statistics. *Theory of Probability & Its Applications*, *49*(3), 377–391.
4. Bening, V. E., & Korolev, V. Yu. (2008). Some statistical problems related to the Laplace distribution (in Russian). *Informatics and Applications*, *2*(2), 19–34.
5. Bening, V. E., Galieva, N. K., & Korolev, V. Yu. (2012). On rate of convergence in distribution of asymptotically normal statistics based on samples of random size. *Annales Mathematicae et Informaticae*, *39*, 17–28.
6. Bening, V. E., Galieva, N. K., & Korolev, V Yu. (2013). Asymptotic expansions for the distribution functions of statistics constructed from samples with random sizes (in Russian). *Informatics and Applications*, *7*(2), 75–83.
7. Buddana, A., & Kozubowski, T. J. (2014). Discrete pareto distributions. *Economic Quality Control*, *29*(2), 143–156.
8. Chen, L. H. Y., Goldstein, L., & Shao, Q.-M. (2011). *Normal approximation by Stein's method*. Probability and its applications. Heidelberg: Springer.
9. Christoph, G., & Wolf, W. (1993). *Convergence theorems with a stable limit law*. Series Mathematical Research. Akademie Verlag.
10. Christoph, G., Ulyanov, V.V., & Bening, V. E. (2019). Second order expansions for sample median with random sample size. arXiv:1905.07765.
11. Christoph, G., Monakhov, M. M., & Ulyanov, V. V. (2020). Second-order Chebyshev–Edgeworth and Cornish–Fisher expansions for distributions of statistics constructed with respect to samples of random size. *Journal of Mathematical Sciences*, *244*(5), 811–839.
12. Cornish, E. A., & Fisher, R. A. (1937). Moments and cumulants in the specification of distributions. *Revue de l'Institut international de Statistique*, *4*, 307–320.
13. Döbler, Ch. (2015). New Berry–Esseen and Wasserstein bounds in the CLT for non-randomly centered random sums by probabilistic methods. *ALEA Latin American Journal of Probability and Mathematical Statistics*, *12*(2), 863–902.

14. Eichelsbacher, P., & Löwe, M. (2019). Lindeberg's method for moderate deviations and random summation. *Journal of Theoretical Probability, 32*(2), 872–897.
15. Esquível, M. L., Mota, P. P., & Mexia, J. T. (2016). On some statistical models with a random number of observations. *Journal of Statistical Theory and Practice, 10*(4), 805–823.
16. Fujikoshi, Y., Ulyanov, V. V., & Shimizu, R. (2010). *Multivariate analysis: High-dimensional and large-sample approximations*. Hoboken: Wiley.
17. Gnedenko, B. V. (1989). An estimate of the distribution of the unknown parameters with a random number of independent observations (in Russian). *Proceedings of Tbilisi Mathematics Institute, AN GSSR, 92*, 146–150.
18. Gnedenko, B. V., & Korolev, V. Yu. (1996). *Random summation*. Limit theorems and applications. Boca Raton: CRC Press.
19. Hill, G. W., & Davis, A. W. (1968). Generalized asymptotic expansions of Cornish–Fisher type. *The Annals of Mathematical Statistics, 39*, 1264–1273.
20. Jaschke, S. (2002). The Cornish-Fisher expansion in the context of delta-gamma-normal approximations. *Journal of Risk, 4*(2), 33–52.
21. Johnson, N. L., Kemp, A. W., & Kotz, S. (2005). *Univariate discrete distributions* (3rd ed.). Hoboken: Wiley.
22. Kalashnikov, V. (1997). *Geometric sums: Bounds for rare events with applications: Risk analysis, reliability, queueing. Mathematics and its applications* (Vol. 413). Dordrecht: Kluwer Academic.
23. Klüppelberg, C., & Mikosch, T. (1997). Large deviations of heavy-tailed random sums with applications in insurance and finance. *Journal of Applied Probability, 34*(2), 293–308.
24. Lyamin, O. O. (2010). On the rate of convergence of the distributions of certain statistics to the Laplace distribution. *Vestnik Moskov University Series XV, 3*, 30–37.
25. Nunes, C., Capistrano, G., Ferreira, D., Ferreira, S. S., & Mexia, J. T. (2019). Exact critical values for one-way fixed effects models with random sample sizes. *Journal of Computational and Applied Mathematics, 354*, 112–122.
26. Peköz, E. A., Röllin, A., & Ross, N. (2013). Total variation error bounds for geometric approximation. *Bernoulli, 19*(2), 610–632.
27. Petrov, V. V. (1995). *Limit theorems of probability theory*. Sequences of independent random variables. Oxford: Clarendon Press.
28. Pike, J., & Ren, H. (2014). Stein's method and the Laplace distribution. *ALEA Latin American Journal of Probability and Mathematical Statistics, 11*(1), 571–587.
29. Robbins, H. (1948). The asymptotic distribution of the sum of a random number of random variables. *Bulletin of the American Mathematical Society, 54*, 1151–1161.
30. Schluter, C., & Trede, M. (2016). Weak convergence to the Student and Laplace distributions. *Journal of Applied Probability, 53*, 121–129.
31. Ulyanov, V. V. (2011). Cornish–Fisher expansions. In M. Lovric (Ed.), *International encyclopedia of statistical science* (pp. 312–315). Berlin: Springer.
32. Ulyanov, V. V., Aoshima, M., & Fujikoshi, Y. (2016). Non-asymptotic results for Cornish–Fisher expansions. *Journal of Mathematical Sciences, 218*(10), 84–91.
33. Wald, A. (1945). Some generalizations of the theory of cumulative sums of random variables. *The Annals of Mathematical Statistics, 16*, 287–293.
34. Wilks, S. S. (1959). Recurrence of extreme observations. *Journal of the Australian Mathematical Society, 1*(1), 106–112.

Chapter 10
Power-Divergence Statistics

Abstract In this chapter, we focus on approximation problems motivated by studies on the asymptotic behavior of power-divergence family of statistics. These statistics are the goodness-of-fit test statistics and include, in particular, the Pearson chi-squared statistic, the Freeman–Tukey statistic, and the log-likelihood ratio statistic. The distributions of the statistics converge to the chi-squared distribution as sample size n tends to ∞. We show that the rate of convergence is of order $n^{-\alpha}$ with $\alpha : 1/2 < \alpha < 1$. Under some conditions α is close to 1. The proofs are based on the fundamental number theory results about approximating the number of integer points in convex sets by the Lebesgue measure of the set.

10.1 Introduction

Take a vector $\boldsymbol{Y} = (Y_1, \ldots, Y_k)^T$ with multinomial distribution $M_k(n, \pi)$, i.e.,

$$\Pr\{Y_1 = n_1, \ldots, Y_k = n_k\} = \begin{cases} n! \prod_{j=1}^k (\pi_j^{n_j}/n_j!), & n_j = 0, 1, \ldots, n\ (j = 1, \ldots, k) \\ & \text{and } \sum_{j=1}^k n_j = n, \\ 0, & \text{otherwise,} \end{cases}$$

where $\boldsymbol{\pi} = (\pi_1, \ldots, \pi_k)^T, \pi_j > 0, \sum_{j=1}^k \pi_j = 1$.

For testing the simple hypothesis $H_0 \colon \boldsymbol{\pi} = \boldsymbol{p}$, ($\boldsymbol{p}$ is a fixed vector), against $H_1 : \boldsymbol{\pi} \neq \boldsymbol{p}$, the following three statistics are commonly used:

(1) Pearson's chi-square statistic

$$t_1(\boldsymbol{Y}) = \sum_{j=1}^k (Y_j - np_j)^2/(np_j);$$

(2) Log-likelihood ratio statistic

$$t_0(\boldsymbol{Y}) = 2 \sum_{j=1}^k Y_j \log\{Y_j/(np_j)\};$$

Y. Fujikoshi and V. V. Ulyanov, *Non-Asymptotic Analysis of Approximations for Multivariate Statistics*, JSS Research Series in Statistics,
https://doi.org/10.1007/978-981-13-2616-5_10

(3) Freeman–Tukey statistic

$$t_{-1/2}(\boldsymbol{Y}) = 4\sum_{j=1}^{k}(\sqrt{Y_j} - \sqrt{np_j})^2,$$

where $\boldsymbol{p} = (p_1, \ldots, p_k)$, $p_j > 0$ for $j = 1, \ldots, k$ and $\sum_{j=1}^{k} p_j = 1$.

All these statistics belong to the power-divergence family of goodness-of-fit test statistics:

$$t_\lambda(\boldsymbol{Y}) = \frac{2}{\lambda(\lambda+1)}\sum_{j=1}^{k} Y_j\left[\left(\frac{Y_j}{np_j}\right)^\lambda - 1\right], \ \lambda \in \mathbb{R}, \tag{10.1}$$

where for $\lambda = 0, -1$, this notation is understood as a result of passage to the limit.

These statistics were first introduced in [2, 12]. Setting $\lambda = 1$, $\lambda = -1/2$ and $\lambda = 0$, we can obtain the chi-squared statistic, the Freeman–Tukey statistic and the log-likelihood ratio statistic, respectively.

It is advisable to use another representation for $t_\lambda(\boldsymbol{Y})$ instead of (10.1). For this purpose, consider the transformation

$$X_j = (Y_j - np_j)/\sqrt{n},\ j = 1, \ldots, k,\ r = k-1,\ \boldsymbol{X} = (X_1, \ldots, X_r)^T.$$

Herein the vector $\boldsymbol{X}$ takes values on the lattice

$$L = \left\{\boldsymbol{x} = (x_1, \ldots, x_r)^T;\ \boldsymbol{x} = \frac{\boldsymbol{m} - n\boldsymbol{p}}{\sqrt{n}},\ \boldsymbol{p} = (p_1, \ldots, p_r)^T,\ \boldsymbol{m} = (n_1, \ldots, n_r)^T\right\},$$

where n_j are nonnegative integers.

The statistic $t_\lambda(\boldsymbol{Y})$ can be expressed as a function of $\boldsymbol{X}$ in the form

$$T_\lambda(\boldsymbol{X}) = \frac{2n}{\lambda(\lambda+1)}\left[\sum_{j=1}^{k} p_j\left(\left(1 + \frac{X_j}{\sqrt{n}p_j}\right)^{\lambda+1} - 1\right)\right]. \tag{10.2}$$

It is known that under null hypothesis H_0 the distributions of all statistics in the family converge to chi-squared distribution with $(k-1)$ degrees of freedom (see, e.g., [2], p. 47). However, more intriguing is the problem of finding the rate of convergence to the limiting distribution.

10.2 Rates of Convergence

To formulate the known results on the rate of convergence we need additional notation.

We call a set $B \subset \mathbb{R}^r$ *an extended convex set* if, for all $l = 1, \ldots, r$, it can be expressed in the form

$$B = \{\boldsymbol{x} = (x_1, \ldots, x_r)^T : \lambda_l(x^*) < x_l < \theta_l(x^*) \text{ and } x^* = (x_1, \ldots, x_{l-1}, x_{l+1}, \ldots, x_r)^T \in B_l\},$$

where $B_l \subset \mathbb{R}^{r-1}$ and $\lambda_l(x^*)$, $\theta_l(x^*)$ are continuous functions on $\mathbb{R}^{r-1}$.

Further, put

$$[h(\boldsymbol{x})]_{\lambda_l(x^*)}^{\theta_l(x^*)} = h(x_1, \ldots, x_{l-1}, \theta_l(x^*), x_{l+1}, \ldots, x_r) - h(x_1, \ldots, x_{l-1}, \lambda_l(x^*), x_{l+1}, \ldots, x_r).$$

For any bounded extended convex set B, an asymptotic expansion was obtained in [15], which in [13] was converted to

$$\Pr\{\boldsymbol{X} \in B\} = J_1 + J_2 + O(n^{-1}). \tag{10.3}$$

with

$$J_1 = \int \cdots \int_B \phi(\boldsymbol{x}) \left\{1 + \tfrac{1}{\sqrt{n}} h_1(\boldsymbol{x}) + \tfrac{1}{n} h_2(\boldsymbol{x})\right\} dx, \text{ where}$$
$$h_1(\boldsymbol{x}) = -\tfrac{1}{2} \sum_{j=1}^k \tfrac{x_j}{p_j} + \tfrac{1}{6} \sum_{j=1}^k x_j \left(\tfrac{x_j}{p_j}\right)^2,$$
$$h_2(\boldsymbol{x}) = \tfrac{1}{2} h_1(\boldsymbol{x})^2 + \tfrac{1}{12} \left(1 - \sum_{j=1}^k \tfrac{1}{p_j}\right) + \tfrac{1}{4} \sum_{j=1}^k \left(\tfrac{x_j}{p_j}\right)^2 - \tfrac{1}{12} \sum_{j=1}^k x_j \left(\tfrac{x_j}{p_j}\right)^3,$$

$$J_2 = -\frac{1}{\sqrt{n}} \sum_{l=1}^r n^{-(r-l)/2} \sum_{x_{l+1} \in L_{l+1}} \cdots \sum_{x_r \in L_r} \left[\int \cdots \int_{B_l} [S_1(\sqrt{n}x_l + np_l)\phi(\boldsymbol{x})]_{\lambda_l(x^*)}^{\theta_l(x^*)} dx_1, \cdots, dx_{l-1}\right], \tag{10.4}$$

$$L_j = \left\{\boldsymbol{x} \colon x_j = \tfrac{n_j - np_j}{\sqrt{n}},\ n_j \text{ and } p_j \text{are defined above}\right\},$$
$$S_1(x) = x - \lfloor x \rfloor - 1/2,\ \lfloor x \rfloor \text{ is the integer part of } x,$$
$$\phi(\boldsymbol{x}) = \tfrac{1}{(2\pi)^{r/2} |\Omega|^{1/2}} \exp\left(-\tfrac{1}{2} \boldsymbol{x}^T \Omega^{-1} \boldsymbol{x}\right).$$

In [15], it was shown that

$$J_2 = O(n^{-1/2}). \tag{10.5}$$

In [15], it was also examined expansion for the most well-known power-divergence statistic, namely, the chi-squared statistic. Set $B^\lambda = \{\boldsymbol{x} : T_\lambda(\boldsymbol{x}) < c\}$. It is straightforward to show that B^1 is an ellipsoid, which is a particular case of a bounded extended convex set. Yarnold managed to simplify the item (10.4) in this case and convert the expansion (10.3) to

$$\Pr\{X \in B^1\} = G_r(c) + (N^1 - n^{r/2}V^1)e^{-c/2} \Big/ \left((2\pi n)^r \prod_{j=1}^{k} p_j\right)^{1/2} + \mathrm{O}(n^{-1}), \tag{10.6}$$

where $G_r(c)$ is the chi-squared distribution function with r degrees of freedom, N^1 is the number of points of the lattice L in B^1, and V^1 is the volume of B^1. Using the well-known result from [3] to estimate the second term in (10.6), Yarnold in [15] proved that

$$\Pr\{T_1(X) < c\} = \Pr\{X \in B^1\} = G_r(c) + \mathrm{O}(n^{-(k-1)/k}). \tag{10.7}$$

In [13], it was shown that, when $\lambda = 0, \lambda = -1/2$, we have

$$\begin{aligned} J_1 &= G_r(c) + \mathrm{O}(n^{-1}), \\ J_2 &= (N^\lambda - n^{r/2}V^\lambda)e^{-c/2} \Big/ \left((2\pi n)^r \textstyle\prod_{j=1}^k p_j\right)^{1/2} + \mathrm{o}(1), \end{aligned} \tag{10.8}$$

where N^λ is the number of points of L in B^λ, and V^λ is the volume of B^λ.

These results were extended by Read to the general case of $\lambda \in \mathbb{R}$. In particular, Theorem 3.1 in [12] implies

$$\Pr\{T_\lambda(X) < c\} = G_r(c) + J_2 + \mathrm{O}\left(n^{-1}\right). \tag{10.9}$$

This reduces the problem to the estimation of the order of J_2.

Note that in [12, 13] there is no estimate for the remainder term in (10.8), just $\mathrm{o}(1)$. Therefore, it is impossible to construct estimates of the rate of convergence of the statistics T_λ to the limiting distribution, based on the simple representation for J_2 initially suggested by Yarnold. Since $B^\lambda = \{x : T_\lambda(x) < c\}$ is a bounded extended convex set, in general case of $\lambda \in \mathbb{R}$, applying (10.5), one has only

$$\Pr(T_\lambda(X) < c) = G_r(c) + \mathrm{O}\left(n^{-1/2}\right). \tag{10.10}$$

However, this result can be significantly improved. We show it in next section.

10.3 Refinements of Convergence Rates

First, estimating the second summand in (10.6) on the base of the result from [4] for the lattice problem for quadratic forms, in the case of Pearson chi-squared statistics, (see [5]) i.e., when $\lambda = 1$, one has for all $r \geq 5$

$$\Pr\{T_1(X) < c\} = \Pr\{X \in B^1\} = G_r(c) + \mathrm{O}(n^{-1}). \tag{10.11}$$

The relation (10.11) not only improves (10.7) but also it is optimal in the sense that it is impossible to get $O(n^{-\alpha})$ with $\alpha > 1$. The extension of (10.11) to the non-discrete random vectors see, e.g., Sect. 1 in [11].

Now, consider the general case of $\lambda \in \mathbb{R}$. In [14] it is shown that $B^\lambda = \{\boldsymbol{x} : T_\lambda(\boldsymbol{x}) < c\}$ is a bounded extended convex (strictly convex) set. The relation (10.8) was improved in [14]:

$$J_2 = (N^\lambda - n^{r/2}V^\lambda)e^{-c/2} \Big/ \left((2\pi n)^r \prod_{j=1}^{k} p_j \right)^{1/2} + O(n^{-1}).$$

Therefore, it is sufficient to use the results from number theory about approximation of the number of integer points in convex sets (more general than ellipsoids) by the Lebesgue measure of the set, see the theorems in next section. Thus, a key issue is to investigate the applicability of the aforementioned theorems from number theory to the set B^λ, see discussion after Theorem 10.4 in next section.

In this way, applying Theorem 10.3, for the specific case of $r = 2$, the estimate (10.5) was considerably refined in [1]:

$$J_2 = O\left(n^{-3/4+\varepsilon}(\log n)^{315/146}\right) \tag{10.12}$$

with $\varepsilon = 3/4 - 50/73 < 0.0651$. According to (10.8), the rate of convergence of J_2 to 0 cannot be better than the results in the lattice point problem for the ellipsoids in number theory, where for $r = 2$ we have the lower bound of the order $O(n^{-3/4} \log\log n)$; see [7]. Thus, for J_2 the relation (10.12) gives a near-optimal order.

Applying Theorem 10.4, for any fixed dimension $r \geqslant 3$, it was proved in [14] the following theorem.

Theorem 10.1 *For the term* J_2, *from* (10.9), *the following estimate holds:*

$$J_2 = O\left(n^{-r/(r+1)}\right), \quad r \geqslant 3,$$

Theorem 10.1, relations (10.9) and (10.12) imply for the statistics $t_\lambda(\boldsymbol{Y})$ and $T_\lambda(\boldsymbol{X})$ (see (10.1) and (10.2)), the following theorem.

Theorem 10.2 *For* $r \geqslant 3$, *one has*

$$\Pr\{T_\lambda(\boldsymbol{X}) < c\} = G_r(c) + O\left(n^{-r/(r+1)}\right),$$

and for $r = 2$,

$$\Pr\{T_\lambda(\boldsymbol{X}) < c\} = G_2(c) + O\left(n^{-50/73}(\log n)^{315/146}\right).$$

It would be interesting to generalize Theorem 10.2 to phi-divergence statistics, see, e.g., Chap. 2 in [10].

10.4 Approximating the Number of Integer Points in Convex Sets

In [1] for the case of $r = 2$ the following theorem from [9] was used:

Theorem 10.3 *Let D be a two-dimensional convex set with area A, bounded by a simple closed curve C, divided into a finite number of pieces each being 3 times continuously differentiable in the sense that, on each piece C_i, the radius of curvature ρ is positive (and not infinite), continuous and continuously differentiable with respect to the angle of contingency ψ. Then in a set obtained from D by translation and linear expansion of order M, the number of integer points equals*

$$N = AM^2 + \mathrm{O}\left(IM^K (\log M)^\Lambda\right),$$
$$K = \tfrac{46}{73},\ \Lambda = \tfrac{315}{146},$$

where I is a number depending only on the properties of the curve C, and not on the parameter M or A.

In [14] the general case of $r \geqslant 3$ was studied. The main reason why two cases are considered separately, i.e., when $r = 2$ and $r \geq 3$, is due to the fact that it is much more difficult to check applicability to the number theory results for B^λ when $r \geq 3$ compared to when $r = 2$. We used the following result in [14] which derives from [8]:

Theorem 10.4 *Let D be a compact convex set in $\mathbb{R}^m$ with the origin as its inner point. We denote the volume of this set by A. Assume that the boundary of this set is an $(m-1)$-dimensional surface of class $\mathbf{C}^\infty$, the Gaussian curvature being non-zero and finite everywhere on the surface. Also assume that a specially defined "canonical" map from the unit sphere to D is one-to-one and belongs to the class $\mathbf{C}^\infty$. Then, in the set that is obtained from the initial set by translation along an arbitrary vector and by linear expansion with factor M, the number of integer points is*

$$N = AM^m + \mathrm{O}\left(IM^{m-2+2/(m+1)}\right), \tag{10.13}$$

where the constant I is a number dependent only on the properties of curve C, and not on parameter M or A.

Provided that $m = 2$, the statement of Theorem 10.4 is weaker than the result of Huxley.

The above theorem is applied in [14] with $M = \sqrt{n}$. Therefore, for any fixed λ, we have to deal not with a single set, but rather with a sequence of sets $B^\lambda(n)$ which are, however, "close" to the limiting set B^1 for all sufficiently large n. In fact, via Taylor expansion, we get

$$T_\lambda(X) = \sum_{i=1}^{k} \left(\frac{X_i^2}{p_i} + \frac{(\lambda-1)X_i^3}{3\sqrt{n}p_i^2} + \frac{(\lambda-1)(\lambda-2)X_i^4}{12p_i^3 n} + \mathrm{O}\left(n^{-3/2}\right)\right).$$

Clearly, the statistic $T_\lambda(\boldsymbol{X})$ is "close" to quadratic form

$$T_1(\boldsymbol{X}) = \sum_{i=1}^{k} \frac{X_i^2}{p_i}.$$

Therefore, $B^\lambda(n)$ is "close" to B^1 for all sufficiently large n.

It is necessary to emphasize that, generally speaking, the constant I in (10.13) in our case is $I(n)$, i.e., it depends on n. Only having ascertained that the inequality

$$|I(n)| \leqslant C_0$$

holds, where C_0 is an absolute constant, we are able to apply Theorem 10.4 without a change of the overall order of the error with respect to n.

We did not use recent results refining Theorems 10.3 and 10.4, see, e.g., Sect. 27 in [6], because their forms of remainder term are not suitable for analysis of properties of $I(n)$.

References

1. Asylbekov, Zh. A., Zubov, V. N., & Ulyanov, V. V. (2011). On approximating some statistics of goodness-of-fit tests in the case of three-dimensional discrete data. *Siberian Mathematical Journal*, *52*(4), 571–584.
2. Cressie, N. A. C., & Read, T. R. C. (1984). Multinomial goodness-of-fit tests. *Journal of the Royal Statistical Society, Series B*, *46*, 440–464.
3. Esseen, C. G. (1945). Fourier analysis of distribution functions. *Acta Mathematica*, *77*, 1–125.
4. Götze, F. (2004). Lattice point problems and values of quadratic forms. *Inventiones Mathematicae*, *157*, 195–226.
5. Götze, F., & Ulyanov, V. V. (2003). *Asymptotic disrtribution of χ^2-type statistics*, Preprint 03-033, Research group "Spectral analysis, asymptotic distributions and stochastic dynamics".
6. Gruber, P. M. (2007). *Convex and discrete geometry*. New York: Springer.
7. Hardy, G. (1916). On Dirichlet's divisor problem. *Proceedings of the London Mathematical Society*, *15*, 1–25.
8. Hlawka, E. (1950). Über integrale auf konvexen körpern. *Monatshefte für Mathematik*, *54*, 1–36.
9. Huxley, M. N. (1993). Exponential sums and lattice points II. *Proceedings of London Mathematical Society*, *66*, 279–301.
10. Pardo, L. (2006). *Statistical inference based on divergence measures*. Boca Raton: Chapman & Hall/CRC.
11. Prokhorov, Y. V., & Ulyanov, V. V. (2013). Some approximation problems in statistics and probability. *Limit theorems in probability, statistics and number theory* (Vol. 42, pp. 235–249)., Springer proceedings in mathematics & statistics Heidelberg: Springer.
12. Read, T. R. C. (1984). Closer asymptotic approximations for the distributions of the power divergence goodness-of-fit statistics. *The Annals of Statistics*, *36*, 59–69.
13. Siotani, M., & Fujikoshi, Y. (1984). Asymptotic approximations for the distributions of multinomial goodness-of-fit statistics. *Hiroshima Mathematical Journal*, *14*, 115–124.

14. Ulyanov, V. V., & Zubov, V. N. (2009). Refinement on the convergence of one family of goodness-of-fit statistics to chi-squared distribution. *Hiroshima Mathematical Journal*, *39*(1), 133–161.
15. Yarnold, J. K. (1972). Asymptotic approximations for the probability that a sum of lattice random vectors lies in a convex set. *The Annals of Mathematical Statistics*, *43*, 1566–1580.

Chapter 11
General Approach to Constructing Non-Asymptotic Bounds

Abstract In this chapter, we consider asymptotic expansions for a class of sequences of symmetric functions of many variables. It implies a general approach to get the non-asymptotic bounds for accuracy of approximation of nonlinear forms in random elements in terms of Lyapunov type ratios. Applications to classical and free probability theory are discussed. In particular, we apply the general results to the central limit theorem for weighted sums, including the case of dependent summands and the case when the distributions of weighted sums are approximated by the normal distribution with accuracy of order $O(n^{-1})$. We consider also applications for distributions of U-statistics of the second order and higher.

11.1 Introduction

Most limit theorems such as the central limit theorem in finite dimensional and abstract spaces and the functional limit theorems admit refinements in terms of asymptotic expansions in powers of $n^{-1/2}$, where n denotes the number of random elements (observations). Results concerning asymptotic expansions of this type are summarized in many monographs; see, e.g., [3].

These expansions are obtained by very different techniques, such as expanding the characteristic function of the particular statistic in the case of linear statistics of independent observations; see, e.g., Chap. 2, in [3] and Chap. 6, in [20]. Other techniques combine convolutions and characteristic functions to develop expansions for quadratic forms (see, e.g., [14, 21]) or for some discrete distributions expansions are derived starting from a combinatorial formula for its distribution function (see, e.g., [7, 18]). Alternatively, one might use an expansion for an underlying empirical process and evaluate it on a domain defined by a functional or statistic of this process. In those cases, one would need to make approximations by Gaussian processes in suitable function spaces.

Y. Fujikoshi and V. V. Ulyanov, *Non-Asymptotic Analysis of Approximations for Multivariate Statistics*, JSS Research Series in Statistics,
https://doi.org/10.1007/978-981-13-2616-5_11

The aim of this chapter is to show that, for most of these expansions, one could safely ignore the underlying probability model and its ingredients (like, e.g., proof of existence of limiting processes and its properties). Indeed, similar expansions can be derived in all of these models by using a general scheme reflecting some (hidden) common features. This is the universal collective behavior caused by many independent asymptotically negligible variables influencing the distribution of a functional.

The results in this chapter can be seen as an extension of the results given by Götze in [9], where the following scheme of sequences of symmetric functions is studied. Let $h_n(\varepsilon, \dots, \varepsilon_n), n \geq 1$ denote a sequence of real functions defined on $\mathbb{R}^n$ and suppose that the following conditions hold:

$$h_{n+1}(\varepsilon_1, \dots, \varepsilon_j, 0, \varepsilon_{j+1}, \dots, \varepsilon_n) = h_n(\varepsilon_1, \dots, \varepsilon_j, \varepsilon_{j+1}, \dots, \varepsilon_n), \tag{11.1}$$

$$\frac{\partial}{\partial \varepsilon_j} h_n(\varepsilon_1, \dots, \varepsilon_j, \dots, \varepsilon_n)\Big|_{\varepsilon_j=0} = 0 \text{ for all } j = 1, \dots, n, \tag{11.2}$$

$$h_n(\varepsilon_{\pi(1)}, \dots, \varepsilon_{\pi(n)}) = h_n(\varepsilon_1, \dots, \varepsilon_n) \text{ for all } \pi \in S_n, \tag{11.3}$$

where S_n denotes a symmetric group.

Consider the following class of examples. Set

$$h_n(\varepsilon_1, \dots, \varepsilon_n) = \mathrm{E}_P(F(\varepsilon_1(\delta_{X_1} - P) + \dots + \varepsilon_n(\delta_{X_n} - P))), \tag{11.4}$$

where F denotes a smooth functional defined on the space of signed measures and X_j denote random elements (in an arbitrary space) with common distribution P. Thus h_n is the expected value of the functional F of a *weighted* empirical process (based on the Dirac measures in $X_1, \dots, X_n$). Here property (11.1) is obvious. Property (11.2), that is, the locally quadratic dependence on weights around zero, is a consequence of the smoothness of F and the centering in P-mean of the signed random measures $\delta_{X_J} - P$ and (11.3) follows from the identical distribution of theX_j's. Properties (11.1) and (11.3) suggest considering the argument ϵ_j of h_n as a weight which controls the effect that X_j has on the distribution of the functional. In [9], Götze considered limits and expansions for functions h_n of equal weights $\varepsilon_j = n^{-1/2}, 1 \leq j \leq n$ with applications to the case (11.4). Those results explained the common structure of expansions for identical weights developed in, e.g., [3, 8, 10, 12]. In the following, this scheme will be extended to the case of *non-identical* weights ε_j, like in the class of examples where the functions h_n are given by (11.4). Moreover, the dependence of F on the elements X_j may be nonlinear.

Denote by ε the n-vector $\varepsilon_j, 1 \leq j \leq n$ and by $\varepsilon^d := \sum_{j=1}^n \varepsilon_j^d, d \geq 1$, the dth power sum. In the following, we show that (11.1)–(11.3) ensure the existence of a "limit" function $h_\infty(\varepsilon^2, \lambda_1, \dots, \lambda_s)$ as a first-order approximation of h_n together with "Edgeworth-type" asymptotic expansions; see, e.g., the case of sums of non-identically distributed random variables in Chap. 6, [20]. These expansions are given in terms of polynomials of power sums $\varepsilon^d, d \geq 3$. The coefficients of these "Edgeworth polynomials", defined in (11.12) below, are given by derivatives of the limit function h_∞ at $\lambda_1 = 0, \dots, \lambda_s = 0$.

Remark 11.1 (*Algebraic Representations*) In the case that h_n is a *multivariate polynomial* of ε satisfying (11.1)–(11.3), we may express it as a polynomial in the algebraic base, ε^d, $d \geq 1$, of symmetric power sums of ε with constant coefficients. Note that

$$\frac{\partial}{\partial_{\varepsilon_j}} \varepsilon^d \bigg|_{\varepsilon_j=0} = \delta_{d,1},$$

where $\delta_{d,1} = 1$ if $d = 1$, and zero otherwise. Hence, (11.2) entails that, in this representation, h_n is a polynomial of ε^d, $d \geq 2$ only and does not depend on ε^1. Now we may write

$$h_n(\varepsilon) = P_{\varepsilon^2}(\varepsilon^3, \ldots, \varepsilon^n),$$

where P_{ε^2} denotes a polynomial with coefficients in the polynomial ring $\mathrm{C}[\varepsilon^2]$ of the variable ε^2. Restricting ourselves to the sphere $\varepsilon^2 = 1$ for convenience, P_{ε^2} is the desired "Edgeworth expansion", provided we introduce the grading of monomials in the variables ε^d, $d \geq 3$ via $\deg(\varepsilon^d) := d - 2$ and expand the polynomial h_n in monomials of $\varepsilon^3, \ldots, \varepsilon^n$ by collecting terms according to this grading.

11.1.1 Notation

Throughout the chapter, the following notation will be used. We define $\varepsilon^d := \sum_{i=1}^n \varepsilon_i^d$ and $|\varepsilon|^d := \sum_{i=1}^n |\varepsilon_i|^d$. Furthermore, we denote by $(\varepsilon)_d$ and $|\varepsilon|_d$ the dth root of ε^d and $|\varepsilon|^d$, respectively, i.e., $(\varepsilon)_d := (\varepsilon^d)^{1/d}$ and $|\varepsilon|_d := (|\varepsilon|^d)^{1/d}$. By c or C, with or without indices, we denote absolute constants, which can be different in different contexts. Let D^α, where α is a nonnegative integral vector, denote partial derivatives $\frac{\partial^{\alpha_1}}{\partial \varepsilon_1^{\alpha_1}} \cdots \frac{\partial^{\alpha_m}}{\partial \varepsilon_m^{\alpha_m}}$, and finally let $\alpha = \sum_{j=1}^m \alpha_j$.

11.2 Results for Symmetric Functions

For an integer $s \geq 0$, introduce the functions

$$h_\infty(\lambda_1, \ldots, \lambda_s, \lambda) := \lim_{k \to \infty} h_{k+s}\left(\lambda_1, \ldots, \lambda_s, \frac{\lambda}{\sqrt{k}}, \ldots, \frac{\lambda}{\sqrt{k}}\right). \tag{11.5}$$

Thus, we consider the limit functions of h_{k+s} as $k \to \infty$, where all but s arguments are equal, asymptotically negligible and taken from a $k - 1$-sphere. In the theorems below, we give sufficient conditions for the existence of the limits.

11.2.1 Limit Theorem with Bound for Remainder Term

The following theorem is an analogue of the Berry–Esseen-type inequality for sums of non-identically distributed independent random variables in probability theory; see, e.g., Chap. 6 in [20].

Theorem 11.1 *Assume that $h_n(\cdot)$, $n \geq 1$, satisfies conditions* (11.1)–(11.3) *and with some positive constant B, we have*

$$|D^{\alpha} h_n(\varepsilon_1, \ldots, \varepsilon_n)| \leq B, \tag{11.6}$$

for all $\varepsilon_1, \ldots, \varepsilon_n$, and for all $\alpha = (\alpha_1, \ldots, \alpha_r)$ with $r \leq 3$ such that

$$\alpha_j \geq 2, \quad j = 1, \ldots, r, \quad \sum_{j=1}^{r} (\alpha_j - 2) \leq 1.$$

Then there exists $h_\infty(|\varepsilon|_2)$ defined by (11.5) *with $s = 0$ and*

$$|h_n(\varepsilon_1, \ldots, \varepsilon_n) - h_\infty(|\varepsilon|_2)| \leq c \cdot B \cdot \max(1, |\varepsilon|_2^3) \cdot |\varepsilon|^3,$$

where c is an absolute constant.

In the case that ε depends on n, this theorem shows that if

$$\lim_{n \to \infty} |\varepsilon|_3 = 0, \tag{11.7}$$

then $h_n(\varepsilon_1, \ldots, \varepsilon_n)$ converges to the limit function $h_\infty(|\varepsilon|_2)$, which depends on $\varepsilon_1, \ldots, \varepsilon_n$ via the l_2-norm $|\varepsilon|_2$ only. This means that the sequence of symmetric functions (invariant with respect to S_n) may be approximated by a rotationally invariant function (invariant with respect to the orthogonal group $\mathcal{O}_n$).

Note though that if (11.7) holds, Theorem 11.1 does not provide an explicit formula for the function $h_\infty(|\varepsilon|_2)$, but guarantees its existence.

Remark 11.2 Investigating distributions of weighted sums, it has been shown in [17, Lemma 4.1] that (11.7) holds with high probability under the uniform measure. See also the inequality (11.18) below.

11.2.2 Ideas of the Proof of Theorem 11.1

We divide the proof into three steps. In the first step, we substitute each argument ε_j by a block of length k of equal variables $\varepsilon_j/\sqrt{k}$. This procedure does not change the l_2-norm $|\varepsilon|_2$. After n steps, we arrive at a function which depends on $n \times k$ arguments:

$$h_{nk}\left(\frac{\varepsilon_1}{\sqrt{k}},\ldots,\frac{\varepsilon_1}{\sqrt{k}},\ldots,\frac{\varepsilon_n}{\sqrt{k}},\ldots,\frac{\varepsilon_n}{\sqrt{k}}\right). \tag{11.8}$$

We show that

$$\left|h_n(\varepsilon_1,\ldots,\varepsilon_n) - h_{nk}\left(\frac{\varepsilon_1}{\sqrt{k}},\ldots,\frac{\varepsilon_1}{\sqrt{k}},\ldots,\frac{\varepsilon_n}{\sqrt{k}},\ldots,\frac{\varepsilon_n}{\sqrt{k}}\right)\right|$$
$$\leq c \cdot B \cdot \max(1, |\varepsilon|_2^3)| \cdot |\varepsilon|^3. \tag{11.9}$$

Hence, this approximation step corresponds to Lindeberg's scheme of replacing the summands in the central limit theorem in probability theory by corresponding Gaussian random variables one by one (see, e.g., [19] and further development in [2] and extension to an invariance principle in [5]). Here the replacement is performed not with a Gaussian variable but with a large block of equal weights of corresponding l_2-norm. In the *second step*, still fixing n, we determine the limit of the sequence of functions (11.8) as k goes to infinity. We show that, in this case, the limit depends on $\varepsilon_1,\ldots,\varepsilon_n$ through its l_2-norm $|\varepsilon|_2$ only:

$$\left|h_{nk}\left(\frac{\varepsilon_1}{\sqrt{k}},\ldots,\frac{\varepsilon_1}{\sqrt{k}},\ldots,\frac{\varepsilon_n}{\sqrt{k}},\ldots,\frac{\varepsilon_n}{\sqrt{k}}\right) - h_k\left(\frac{|\varepsilon|_2}{\sqrt{k}},\ldots,\frac{|\varepsilon|_2}{\sqrt{k}}\right)\right| \leq c(\varepsilon)\cdot B\cdot k^{-1/2}, \tag{11.10}$$

where $c(\varepsilon)$ is some positive constant depending on ε only.

Finally, we may apply the arguments from Proposition 2.1 in [9]. We show that there exists some function $h_\infty(|\varepsilon|_2)$ such that

$$\left|h_k\left(\frac{|\varepsilon|_2}{\sqrt{k}},\ldots,\frac{|\varepsilon|_2}{\sqrt{k}}\right) - h_\infty(|\varepsilon|_2)\right| \leq c(\varepsilon)\cdot B\cdot k^{-1/2}. \tag{11.11}$$

From (11.9)–(11.11), it follows that

$$|h_n(\varepsilon_1,\ldots,\varepsilon_n) - h_\infty(|\varepsilon|_2)| \leq C\cdot B\cdot \max(1, |\varepsilon|_2^3)|\varepsilon|^3 + c(\varepsilon)\cdot B\cdot k^{-1/2}.$$

Taking the limit $k \to \infty$, we conclude the statement of the theorem.
See detailed proof in [15].

11.2.3 Asymptotic Expansions

To formulate the asymptotic expansion of the function $h_n(\cdot)$, $n \geq 1$, we have to introduce additional notation. We introduce the following differential operators by means of formal power series identities. Define cumulant differential operators $\kappa_p(D)$ by means of

$$\sum_{p=2}^{\infty} p!^{-1}\varepsilon^p \kappa_p(D) = \ln\left(1+\sum_{p=2}^{\infty} p!^{-1}\varepsilon^p D^p\right)$$

in the formal variable ε. One may easily compute the first cumulants. For example, $\kappa_2 = D^2$, $\kappa_3 = D^3$, $\kappa_4 = D^4 - 3D^2D^2$. Define Edgeworth polynomials by means of the following formal series in κ_r, τ_r and the formal variable ε:

$$\sum_{r=0}^{\infty} \varepsilon^r P_r(\tau_* \kappa_*) = \exp\left(\sum_{r=3}^{\infty} r!^{-1}\varepsilon^{r-2}\kappa_r\tau_r\right),$$

which yields

$$P_r(\tau_*\kappa_*) = \sum_{m=1}^{r} m!^{-1} \sum_{j_1,\dots,j_m} (j_1+2)!^{-1}\tau_{j_1+2}\kappa_{j_1+2} \times (j_2+2)!^{-1}\tau_{j_2+2}\kappa_{j_2+2}\dots(j_m+2)!^{-1}\tau_{j_m+2}\kappa_{j_m+2}, \tag{11.12}$$

where the sum $\sum_{j_1,\dots,j_m}$ extends over all m-tuples of positive integers $(j_1,\dots,j_m)$ satisfying $\sum_{q=1}^{m} j_q = r$ and $\kappa_* = (\kappa_3,\dots,\kappa_{r+2})$, $\tau_* = (\tau_3,\dots,\tau_{r+2})$. For example,

$$P_1(\tau_*\kappa_*) = \frac{1}{6}\tau_3\kappa_3 = \frac{1}{6}\tau_3^3 D^3, \tag{11.13}$$

$$P_2(\tau_*\kappa_*) = \frac{1}{24}\tau_4\kappa_4 + \frac{1}{72}\tau_3^2\kappa_3\kappa_3 = \frac{1}{24}\tau_4(D^4 - 3D^2D^2) + \frac{1}{72}\tau_3^2 D^3 D^3.$$

In the following theorem, we assume that ε is a vector on the unit sphere, i.e., $|\varepsilon|_2 = 1$. It is also possible to consider the general case $|\varepsilon|_2 = r$, $r > 1$, but then the remainder terms will have a more difficult structure. In what follows, we shall drop the dependence of h_∞ on the argument $|\varepsilon|_2$ in the notation of this function.

Theorem 11.2 *Assume that $h_n(\varepsilon_1,\dots,\varepsilon_n)$, $n \geq 1$, satisfies conditions* (11.1)–(11.3) *together with $|\varepsilon|_2 = 1$. Suppose that*

$$|D^\alpha h_n(\varepsilon_1,\dots,\varepsilon_n)| \leq B, \tag{11.14}$$

for all $\varepsilon_1,\dots,\varepsilon_n$, where B denotes some positive constant, $\alpha = (\alpha_1,\dots,\alpha_r)$, $r \leq s$, and

$$\alpha_j \geq 2, \quad j = 1,\dots,r, \quad \sum_{j=1}^{r}(\alpha_j - 2) \leq s-2.$$

Then

$$h_n(\varepsilon_1,\dots,\varepsilon_n) = h_\infty + \sum_{l=1}^{s-3} P_l(\varepsilon^*\kappa_*)h_\infty(\lambda_1,\dots,\lambda_l)\big|_{\lambda_1=\dots=\lambda_l=0} + R_s,$$

where $P_l(\varepsilon^*\kappa_*)$ *is defined in* (11.12) *with* $\varepsilon^* = (\varepsilon^3, \ldots, \varepsilon^{l+2})$, $\kappa_* = (\kappa_3, \ldots, \kappa_{l+2})$ *and*

$$|R_s| \le c_s \cdot B \cdot |\varepsilon|^s$$

with some absolute constant c_s.

As an example, consider the case $s = 5$. Then, by (11.14),

$$h_n(\varepsilon_1, \ldots, \varepsilon_n) = h_\infty + \frac{\varepsilon^3}{6} \frac{\partial^3}{\partial \lambda^3} h_\infty(\lambda)\big|_{\lambda=0}$$
$$+ \left[\frac{\varepsilon^4}{24} \left(\frac{\partial^4}{\partial \lambda_1^4} - 3 \frac{\partial^2}{\partial \lambda_1^2} \frac{\partial^2}{\partial \lambda_2^2} \right) + \frac{(\varepsilon^3)^2}{72} \frac{\partial^3}{\partial \lambda_1^3} \frac{\partial^3}{\partial \lambda_2^3} \right] h_\infty(\lambda_1, \lambda_2)\big|_{\lambda_1=0, \lambda_2=0}$$
$$+ O(|\varepsilon|^5).$$

See proof of Theorem 11.2 in [15].

11.3 Applications in Probability and Statistics

In this section, we use examples to illustrate how one may apply Theorem 11.2 to derive an asymptotic expansion of various functions in probability theory.

11.3.1 Expansion in the Central Limit Theorem for Weighted Sums

As the first example, let us consider the sequence of independent random variables $X, X_j, j \in \mathbb{N}$, taking values in $\mathbb{R}$ with a common distribution function F. Suppose $E(X) = 0$, $E(X^2) = 1$. Consider the weighted sum $S_\varepsilon = \varepsilon_1 X_1 + \cdots + \varepsilon_n X_n$. As h_n, we may choose the characteristic function of S_ε, i.e.,

$$h_n(\varepsilon_1, \ldots, \varepsilon_n) = E(e^{it(\varepsilon_1 X_1 + \cdots + \varepsilon_n X_n)}).$$

From Theorem 11.1, we know that $h_\infty(|\varepsilon|_2)$ exists provided that the condition (11.6) holds. In our setting, this condition holds when $E|X|^3 < \infty$. It is well known (see, e.g., Chap. 5 in [20]) that

$$h_\infty(|\varepsilon|_2) = E(e^{itG}),$$

where $G \sim N(0, |\varepsilon|_2)$. In what follows, we shall assume $|\varepsilon|_2 = 1$. The rate of convergence is given by $|\varepsilon|_3^3$. If ε is well spread, for example, when $\varepsilon_j = n^{-1/2}$ for all $1 \le j \le n$, then

$$|h_n(\varepsilon_1, \ldots, \varepsilon_n) - h_\infty(|\varepsilon|_2)| \le C \cdot |t|^3 \cdot \frac{\mathrm{E}(|X|^3)}{\sqrt{n}}. \tag{11.15}$$

Of course, this bound does not hold for all $\varepsilon = (\varepsilon_1, \ldots, \varepsilon_n)$ on the unit sphere $S^{n-1} = \{\varepsilon : |\varepsilon|_2 = 1\}$. Consider a simple counterexample. Let $X \sim \mathrm{Uniform}([-\sqrt{3}, \sqrt{3}])$ and $\varepsilon = e_1$. Then $S_\varepsilon = X_1 \sim \mathrm{Uniform}([-\sqrt{3}, \sqrt{3}])$, which is not Gaussian as $n \to \infty$.

Concerning expansions for weighted linear forms, results of [17] imply that the left-hand side of (11.15) has order $\mathrm{O}(1/n)$ for a "large" set of unit vectors ε. The size of this set is measured according to the uniform probability measure, say σ_{n-1}, on the unit sphere S^{n-1}.

Let us now construct an asymptotic expansion using Theorem 11.2. We have, for any integer $s \ge 0$,

$$h_\infty(\lambda_1, \ldots, \lambda_s) = \mathrm{E}(e^{it(\lambda_1 X_1 + \cdots + \lambda_s X_s + G)}).$$

Taking derivatives with respect to $\lambda_1, \ldots, \lambda_s$ at zero and assuming $s \le 2$ as an example, we get

$$\begin{aligned}
\frac{\partial^3}{\partial \lambda_1^3} h_\infty(\lambda_1)\Big|_{\lambda_1=0} &= (it)^3 e^{-t^2/2} \beta_3, \\
\frac{\partial^4}{\partial \lambda_1^4} h_\infty(\lambda_1)\Big|_{\lambda_1=0} &= (it)^4 e^{-t^2/2} \beta_4, \\
\frac{\partial^4}{\partial \lambda_1^2 \partial \lambda_2^2} h_\infty(\lambda_1, \lambda_2)\Big|_{\lambda_1=0, \lambda_2=0} &= (it)^4 e^{-t^2/2} \beta_2^2, \\
\frac{\partial^6}{\partial \lambda_1^3 \partial \lambda_2^3} h_\infty(\lambda_1, \lambda_2)\Big|_{\lambda_1=0, \lambda_2=0} &= (it)^6 e^{-t^2/2} \beta_3^2,
\end{aligned}$$

where $\beta_2 = \mathrm{E}(X^2) = 1$, $\beta_3 = \mathrm{E}(X^3)$, and $\beta_4 = \mathrm{E}(X^4)$. Substituting these equations to (11.15), one has

$$\begin{aligned}
h_n(\varepsilon_1, \ldots, \varepsilon_n) = \mathrm{E}(e^{itG}) &+ \frac{\varepsilon^3}{6}(it)^3 e^{-t^2/2} \beta_3 \\
&+ \frac{\varepsilon^4}{24}[\beta_4 - 3](it)^4 e^{-t^2/2} + \frac{(\beta_3 \varepsilon^3)^2}{72}(it)^6 e^{-t^2/2} + R_5.
\end{aligned}$$

The expansion coincides with the well-known Edgeworth expansion (involving cumulants) for characteristic functions of sums of random variables; see, e.g., Sect. 1, Chap. 6, in [20]. It also coincides with Edgeworth expansions for expectations of smooth functions of sums of random vectors in Euclidean (resp., Banach spaces); see, e.g., [12] (resp., [10]).

Let us concentrate now on the so-called short asymptotic expansion

$$h_n(\varepsilon_1, \ldots, \varepsilon_n) = \mathrm{E}(e^{itG}) + \frac{\varepsilon^3}{6}(it)^3 e^{-t^2/2}\beta_3 + R_4, \tag{11.16}$$

where

$$|R_4| \le C \cdot |t|^4 \cdot \sum_{k=1}^{n} \varepsilon_k^4.$$

It follows from [17, Lemma 4.1] that, for some constants C_1 and C_2 and for all $\rho : 1 > \rho > \exp(-C_1\, n)$, there exists a subset $B \subset S^{n-1}$ such that, for any $\varepsilon \in B$, one has

$$\left|\sum_{k=1}^{n} \varepsilon_k^3\right| + \sum_{k=1}^{n} \varepsilon_k^4 \le \left(\log \frac{1}{\rho}\right)^2 \frac{C_2}{n} \tag{11.17}$$

and $\sigma_{n-1}(B) \ge 1 - \rho$ for the uniform probability measure σ_{n-1} on the unit sphere S^{n-1}.

Thus, combining (11.16) and (11.17), we get, for any $\rho : 1 > \rho > 0$ and all $\varepsilon \in B$,

$$|h_n(\varepsilon_1, \ldots, \varepsilon_n) - \mathrm{E}(e^{itG})| \le C\,(|t|^3 + t^4) \left(\log \frac{1}{\rho}\right)^2 \frac{\beta_4}{n} \tag{11.18}$$

for some constant C (cf. (11.15)).

This property may be generalized to arbitrary functions $h_n(\varepsilon_1, \ldots, \varepsilon_n)$ which satisfy the conditions of Theorem 11.2.

Extending this example, it is possible to apply our result for asymptotic expansion in the central limit theorem for quadratic forms in sums of random elements with values in a Hilbert space including infinite-dimensional cases, see, e.g., [3, 11, 14, 21–23].

Moreover, our result could be helpful in studies of asymptotic expansions for the functionals of weighted sums of *dependent* random variables.

Let $X_1, \ldots, X_n$ be identically distributed symmetric random variables and $\delta_1, \ldots, \delta_n$ be independent Rademacher random variables; i.e., δ_i takes values 1 and -1 with probabilities 1/2. Assume that $\delta_1, \ldots, \delta_n$ are independent of $X_1, \ldots, X_n$. We emphasize that here that it is not necessary for $X_1, \ldots, X_n$ to be independent. To construct asymptotic expansions for $\mathrm{E}(F(\varepsilon_1 X_1 + \cdots + \varepsilon_n X_n))$ with some smooth measurable function F, note that

$$F(\varepsilon_1\, X_1 + \cdots + \varepsilon_n\, X_n) \stackrel{d}{=} F(\varepsilon_1\, \delta_1\, X_1 + \cdots + \varepsilon_n\, \delta_n\, X_n),$$

where $\stackrel{d}{=}$ denotes equality in distribution. Consider functions

$$h_n(\varepsilon_1, \ldots, \varepsilon_n) = \mathrm{E}(F(\varepsilon_1\, \delta_1\, X_1 + \cdots + \varepsilon_n\, \delta_n\, X_n)).$$

The function h_n satisfies the conditions (11.1)–(11.3) provided F is sufficiently smooth.

For instance, we can take, for $i = 1, \dots, n$,

$$X_i = \frac{Y_i}{\sqrt{\varepsilon_1^2 Y_1^2 + \cdots + \varepsilon_n^2 Y_n^2}},$$

where $Y_1, \dots, Y_n$ are independent random variables with a common symmetric distribution. Then $F(\varepsilon_1 X_1 + \cdots + \varepsilon_n X_n)$ is a function of a self-normalized weighted sum; see, e.g., [16].

On the other hand, in the special case $\varepsilon_1 = \cdots = \varepsilon_n = 1/\sqrt{n}$, we may consider $F(X_1/\sqrt{n} + \cdots + X_n/\sqrt{n})$ as a function of exchangeable random variables; see, e.g., [4].

11.3.2 Expansion in the Free Central Limit Theorem

It has been shown recently in [13] that one may apply the results of Theorem 11.2 in the setting of free probability theory.

Denote by $\mathcal{M}$ the family of all Borel probability measures defined on the real line $\mathbb{R}$. Let $X_1, X_2, \dots$ be free self-adjoint identically distributed random variables with distribution $\mu \in \mathcal{M}$. We assume that μ has zero mean and unit variance. Let μ_n be the distribution of the normalized sum $S_n := \frac{1}{\sqrt{n}} \sum_{j=1}^n X_j$. In free probability, the sequence of measures μ_n converges to Wigner's semicircle law ω. Moreover, μ_n is absolutely continuous with respect to the Lebesgue measure for sufficiently large n. We denote by p_{μ_n} the density of μ_n. Define the Cauchy transform of a measure μ:

$$G_\mu(z) = \int_{\mathbb{R}} \frac{\mu(dx)}{z - x}, \quad z \in \mathbb{C}^+,$$

where $\mathbb{C}^+$ denotes the upper half-plane.

In [6], Chistyakov and Götze obtained a formal power expansion for the Cauchy transform of μ_n and Edgeworth-type expansions for μ_n and p_{μ_n}. In [13], the general scheme from [9] was applied to derive a similar result.

11.3.3 Expansion of Quadratic von Mises Statistics

Let $X, \overline{X}, X_1, \dots, X_n$ be independent identically distributed random elements taking values in an arbitrary measurable space $(\mathcal{X}, \mathcal{B})$. Assume that $g : \mathcal{X} \to \mathbb{R}$ and $h : \mathcal{X} \times \mathcal{X} \to \mathbb{R}$ are real-valued measurable functions. In addition, we assume that h is symmetric. We consider the quadratic functional

$$w_n(\varepsilon_1, \ldots, \varepsilon_n) = \sum_{k=1}^{n} \varepsilon_j g(X_j) + \sum_{j,k=1}^{n} \varepsilon_j \varepsilon_k h(X_j, X_k),$$

assuming that $\mathrm{E}(g(X)) = 0, \quad \mathrm{E}(h(X, \overline{X})|X) = 0$. We shall derive an asymptotic expansion of

$$h_n(\varepsilon_1, \ldots, \varepsilon_n) := \mathrm{E}(\exp(itw_n(\varepsilon_1, \ldots, \varepsilon_n))).$$

Consider the measurable space $(\mathcal{X}, \mathcal{B}, \mu)$ with measure $\mu := \mathcal{L}(X)$. Let $L^2 := L^2(\mathcal{X}, \mathcal{B}, \mu)$ denote the real Hilbert space of square integrable real functions. A Hilbert–Schmidt operator $\mathbb{Q} : L^2 \to L^2$ is defined via

$$\mathbb{Q}f(x) = \int_{\mathcal{X}} h(x, y) f(y) \mu(dy) = \mathrm{E}(h(x, X) f(X)), \quad f \in L^2.$$

Let $\{e_j, j \geq 1\}$ denote a complete orthonormal system of eigenfunctions of $\mathbb{Q}$ ordered by decreasing absolute values of the corresponding eigenvalues $q_1, q_2, \ldots$, that is, $|q_1| \geq |q_2| \geq \ldots$. Then

$$\mathrm{E}(h^2(X, \overline{X})) = \sum_{j=1}^{\infty} q_j^2 < \infty, \quad h(x, y) = \sum_{j=1}^{\infty} q_j e_j(x) e_j(y).$$

If the closed span $\langle \{e_j, j \geq 1\} \rangle \subset L^2$ is a proper subset, it might be necessary to choose functions e_{-1}, e_0 such that $\{e_j, j = -1, 0, 1, \ldots\}$ is an orthonormal system and

$$g(x) = \sum_{k=0}^{\infty} g_k e_k(x), \quad h(x, x) = \sum_{k=-1}^{\infty} h_k e_k(x).$$

It is straightforward that $\mathrm{E}(e_j(X)) = 0$ for all j. Therefore, $\{e_j(X), j = -1, 0, 1, \ldots\}$ is an orthonormal system of mean zero random variables.

We derive an expression for the derivatives of $h_\infty(\lambda_1, \ldots, \lambda_r)$. Since for every fixed k, the sum $n^{-1/2}(e_k(X_1) + \cdots + e_k(X_n))$ weakly converges to a standard normal random variable, we conclude that $w_{n+r}(\lambda_1, \ldots, \lambda_r, n^{-1/2}, \ldots, n^{-1/2})$ weakly converges to the random variable

$$\begin{aligned} w_\infty(\lambda_1, \ldots, \lambda_r) := w_r(\lambda_1, \ldots, \lambda_r) + \sum_{k=0}^{\infty} g_k Y_k + \sum_{k=1}^{\infty} q_k^2 (Y_k^2 - 1) \\ + \mathrm{E}(h(X, X)) + 2 \sum_{k=1}^{\infty} q_k \left(\sum_{l=1}^{r} \lambda_l e_k(X_l) \right) Y_k, \end{aligned}$$

where $Y_k, k \geq 0$, are independent standard normal random variables. For every fixed T, we get by complex integration

$$\mathrm{E}\left(\exp\left[itq_k(Y_k^2-1)+2itTY_k\right]\right)=\frac{1}{\sqrt{1-2itq_k}}\exp\left[-itq_k-\frac{2t^2T^2}{\sqrt{1-2itq_k}}\right].$$

This yields

$$h_\infty(\lambda_1,\ldots,\lambda_r)=\varphi(t)\mathrm{E}\big(\exp[itw_r(\lambda_1,\ldots,\lambda_r) + (it)^2\sum_{k=1}^{\infty}q_kT_k(\lambda)(2q_kT_k(\lambda)+g_k)(1-2itq_k)^{-1}]\big), \tag{11.19}$$

where $T_k(\lambda)=\sum_{l=1}^{r}\lambda_l e_k(X_l)$ and

$$\varphi(t)=\left[\prod_{k=1}^{\infty}\frac{1}{\sqrt{1-2itq_k}}\exp(-itq_k)\right] \times\exp\left[it\mathrm{E}(h(X_1,X_1))-\frac{t^2}{2}\sum_{k=0}^{\infty}g_k^2(1-2itq_k)^{-1}\right].$$

Let us introduce the following functions of X and $\overline{X}$:

$$h_t(X,\overline{X}):=h(X,\overline{X})+2it\sum_{k=1}^{\infty}q_k^2e_k(X)e_k(\overline{X})(1-2itq_k)^{-1},$$
$$g_t(X):=g(X)+it\mathrm{E}\big(h_t(X,\overline{X})g(\overline{X})|X\big).$$

Applying this notation, we may rewrite (11.19) in the following way:

$$h_\infty(\lambda_1,\ldots,\lambda_r)=\varphi(t)\mathrm{E}\left(\exp\left[it\sum_{j,k=1}^{n}h_t(X_j,X_k)\lambda_j\lambda_k+it\sum_{j=1}^{r}\lambda_jg_t(X_j)\right]\right).$$

Taking derivatives of h_∞ with respect to $\lambda_1,\ldots,\lambda_r$ at zero, we get

$$h_n(\varepsilon_1,\ldots,\varepsilon_n)=\varphi(t)\sum_{r=0}^{s-3}a_r(t,h,g)+R_s,$$

where

$$a_r(t,h,g) := P_r(\varepsilon^* \kappa_*)$$
$$\times \mathrm{E}\left(\exp\left[it \sum_{j,k=1}^{n} h_t(X_j, X_k)\lambda_j\lambda_k + it \sum_{j=1}^{r} \lambda_j g_t(X_j) \right]\Bigg|_{\lambda_1=\cdots=\lambda_r=0} \right).$$

Higher order U-statistics may be treated by similar arguments. See, for example, [1, 10].

11.3.4 Expansions for Weighted One-Sided Kolmogorov–Smirnov Statistics

Let $X_1, \ldots X_n$ be an independent identically distributed random variable with uniform distribution in [0, 1]. Consider the following statistic: $D^+(\varepsilon_1, \ldots, \varepsilon_n, t) = \sum_{j=1}^{n} \varepsilon_j (\mathbb{I}(X_j \le t) - t)$, where $\mathbb{I}(A)$ is indicator of A. For example, if $\varepsilon_j = n^{-1/2}$, $j = 1, \ldots, n$, then we have $D^+(t) = n^{1/2}(F_n(t) - t)$, where $F_n(t)$ denotes the empirical distribution function of $X_1, \ldots, X_n$. We are interested in the asymptotic expansion of

$$\Pr\{\sup_{0 \le t \le 1} D^+(\varepsilon_1, \ldots, \varepsilon_n, t) > a\}, \quad a > 0.$$

It is well known that $h_\infty(0) = \exp[-2a^2]$ and

$$\begin{aligned} h_\infty(\lambda) &= \int_0^1 \Pr\{x(t) + \lambda(\mathbb{I}(s < t) - t) > a, 0 \le t \le 1\}\, ds \\ &= \int_0^1 \mathrm{E} f_a(s, x(s), \lambda) f_a(1 - s, x(s), -\lambda)\, ds, \end{aligned}$$

where $f_a(s, x, \lambda) = \Pr\{x(t) > a + \lambda t, 0 \le t \le s | x(s) = x\} = \exp(-2a(a + \lambda s - x)/s)$ and $x(t), 0 \le t \le 1$ is a Brownian bridge. For more details, see [9]. Then it follows from Theorem 11.2 that

$$\Pr\{\sup_{0 \le t \le 1} D^+(\varepsilon_1, \ldots, \varepsilon_n, t) > a\} = \left[1 + \frac{1}{6}\varepsilon^3 \frac{\partial}{\partial a} + \mathrm{O}(|\varepsilon|_4^4) \right] \exp(-2a^2).$$

Such expansions for equal weights have been derived, for example, by combinatorial and analytic techniques in [7, 18].

References

1. Bentkus, V., & Götze, F. (1999). Optimal bounds in non-Gaussian limit theorems for U-statistics. *The Annals of Probability*, *27*(1), 454–521.
2. Bergström, H. (1944). On the central limit theorem. *Scandinavian Actuarial Journal*, *1944*(3–4), 139–153.
3. Bhattacharya, R. N., & Rao, R. R. (2010). *Normal approximation and asymptotic expansions*. SIAM-Society for Industrial and Applied Mathematics.
4. Bloznelis, M., & Götze, F. (2002). An Edgeworth expansion for symmetric finite population statistics. *The Annals of Probability*, *30*(3), 1238–1265.
5. Chaterjee, S. (2006). A generalization of the Lindeberg principle. *The Annals of Probability*, *34*(6), 2061–2076.
6. Chistyakov, G. P., & Götze, F. (2013). Asymptotic expansions in the CLT in free probability. *Probability Theory and Related Fields*, *157*(1–2), 107–156.
7. Gnedenko, B. V., Koroluk, V. S., & Skorohod, A. V. (1960). Asymptotic expansions in probability theory. In *Proceedings of 4th Berkeley Symposium on Mathematical Statistics and Probability* (Vol. II, pp. 153–170). Berkeley,: University of California Press.
8. Götze, F. (1981). On Edgeworth expansions in Banach spaces. *The Annals of Probability*, *9*(5), 852–859.
9. Götze, F. (1985). Asymptotic expansions in functional limit theorems. *Journal of Multivariate analysis*, *16*, 1–20.
10. Götze, F. (1989). Edgeworth expansions in functional limit theorems. *The Annals of Probability*, *17*(4), 1602–1634.
11. Götze, F. (2004). Lattice point problems and values of quadratic forms. *Inventiones Mathematicae*, *157*, 195–226.
12. Götze, F., & Hipp, C. (1978). Asymptotic expansions in the central limit theorem under moment conditions. *Zeitschrift für Wahrscheinlichkeitstheorie und Verwandte Gebiete*, *42*(1), 67–87.
13. Götze, F., & Reshetenko, A. (2014). Asymptotic expansions in free limit theorems. arXiv:1408.1360.
14. Götze, F., & Zaitsev, A Yu. (2014). Explicit rates of approximation in the CLT for quadratic forms. *The Annals of Probability*, *42*(1), 354–397.
15. Götze, F., Naumov, A. A., & Ulyanov, V. V. (2017). Asymptotic analysis of symmetric functions. *Journal of Theoretical Probability*, *30*(3), 876–897.
16. Jing, B.-Y., & Wang, Q. (2010). A unified approach to Edgeworth expansions for a general class of statistics. *Statistica Sinica*, *20*, 613–636.
17. Klartag, B., & Sodin, S. (2012). Variations on the Berry–Esseen theorem. *Theory of Probability & Its Applications*, *56*(3), 403–419.
18. Lauwerier, H. A. (1963). The asymptotic expansion of the statistical distribution of N. V. Smirnov. *Zeitschrift für Wahrscheinlichkeitstheorie und Verwandte Gebiete*, *2*, 61–68.
19. Lindeberg, J. W. (1922). Eine neue Herleitung des Exponentialgesetzes in der Wahrscheinlichkeitsrechnung. *Mathematische Zeitschrift*, *15*, 211–225.
20. Petrov, V. V. (1975). *Sums of independent random variables*. New York: Springer.
21. Prokhorov, Y. V., & Ulyanov, V. V. (2013). Some approximation problems in statistics and probability. *Limit theorems in probability, statistics and number theory* (Vol. 42, pp. 235–249)., Springer proceedings in mathematics & statistics Heidelberg: Springer.
22. Ulyanov, V. V. (1986). Normal approximation for sums of nonidentically distributed random variables in Hilbert spaces. *Acta Scientiarum Mathematicarum*, *50*(3–4), 411–419.
23. Ulyanov, V. V. (1987). Asymptotic expansions for distributions of sums of independent random variables in H. *Theory of Probability and its Applications*, *31*(1), 25–39.

Printed in the United States
By Bookmasters